The Gold Dust Beneath Hypercomplex Cosmology
Fractional Creation Operators, Broken Scaling Precursor Model, Universe Computers, Negative Space-times, Dust Confinement

Stephen Blaha Ph. D.
Blaha Research

Pingree Hill Publishing
MMXXII

Rev. 00/00/01   February 15, 2022

To Margaret

Some Other Books by Stephen Blaha

All the Megaverse! Starships Exploring the Endless Universes of the Cosmos using the Baryonic Force (Blaha Research, Auburn, NH, 2014)

SuperCivilizations: Civilizations as Superorganisms (McMann-Fisher Publishing, Auburn, NH, 2010)

All the Universe! Faster Than Light Tachyon Quark Starships & Particle Accelerators with the LHC as a Prototype Starship Drive Scientific Edition (Pingree-Hill Publishing, Auburn, NH, 2011).

Unification of God Theory and Unified SuperStandard Model THIRD EDITION (Pingree Hill Publishing, Auburn, NH, 2018).

The Exact QED Calculation of the Fine Structure Constant Implies ALL 4D Universes have the Same Physics/Life Prospects (Pingree Hill Publishing, Auburn, NH, 2019).

Unified SuperStandard Theory and the SuperUniverse Model: The Foundation of Science (Pingree Hill Publishing, Auburn, NH, 2018).

Quaternion Unified SuperStandard Theory (The QUeST) and Megaverse Octonion SuperStandard Theory (MOST) (Pingree Hill Publishing, Auburn, NH, 2020).

Unified SuperStandard Theories for Quaternion Universes & The Octonion Megaverse (Pingree Hill Publishing, Auburn, NH, 2020).

The Essence of Eternity: Quaternion & Octonion SuperStandard Theories (Pingree Hill Publishing, Auburn, NH, 2020).

From Octonion Cosmology to the Unified SuperStandard Theory of Particles (Pingree Hill Publishing, Auburn, NH, 2020).

Beyond Octonion Cosmology (Pingree Hill Publishing, Auburn, NH, 2021).

Integration of General Relativity and Quantum Theory: Octonion Cosmology, GiFT, Creation/Annihilation Spaces CASe, Reduction of Spaces to a Few Fermions and Symmetries in Fundamental Frames (Pingree Hill Publishing, Auburn, NH, 2021).

Available on Amazon.com, bn.com Amazon.co.uk and other international web sites as well as at better bookstores (through Ingram Distributors).

CONTENTS

FIGURES and TABLES

Introduction

This author created a cosmology using hypercomplex numbers in January, 2020. In the following two years he rapidly evolved it to Octonion Cosmology. Recently he created a new formulation called Hypercomplex Cosmology. This formulation has a much cleaner spectrum of spaces, and it supports a deeper stratum of infinitesimal quantities, Gold Dust,[1] that accretes to form the particles, spaces, masses, and energies of Hypercomplex Cosmology.

This book extends the 10 space spectrum to an infinite number of spaces. The new spectrum is then reduced to the physical 10 space spectrum plus an infinite spectrum of fractional spaces. The additional fractional spaces are used to fractionate universes, particles, interactions, and creation/annihilation operators through a series succession of fineness levels to infinitesimal parts that are called Gold Dust. They are the stuff of everything in Hypercomplex Cosmology. The theory reaches to the level of one substance of varying attributes—realizing the dream of the PreSocratic Philosophers: Anaximander, Thales, Parmenides, Zeno, Democritus, …

Hypercomplex Cosmology is shown to have a basis in scaling using a new model that generates Cayley numbers and the spectrum of 10 spaces. A breakdown of scaling limits the number of physical spaces to 10.

The extension to the consideration of an infinite set of spaces leads to negative space-time dimensions, which are physically reasonable.

Hypercomplex Cosmology is shown to be based on infinitesimal local parts, Gold Dust, that combine to form the particles of Nature with Gold Dust confinement analogous to quark confinement.

The critical importance of creation/annihilation operators in the formulation of Hypercomplex Cosmology led the author to revive his 1998 formulation of the Cosmos as a computer program in a computer language defined with creation/annihilation operators. Hypercomplex Cosmology has a clear computation formulation.

This book is best understood if the reader is familiar with the author's previous book *New View: Hypercomplex Cosmology Based on the Unification of General Relativity and Quantum Theory*.

---

[1] Signifying an auspicious beginning for the Cosmos.

1. Hypercomplex Cosmology

We have developed Hypercomplex Cosmology, first in the form of Octonion Cosmology, and secondly, in view of perceived shortcomings by the author, in its new form in our recent books. The new spectrum of spaces is cleaner, mathematically and physically, then that of Octonion Cosmology. See Fig. 1.1.

There are a number of issues with the spectrum that require resolution: Why the upper limit of space n = 10? Why not larger spaces? Why the self-similar scaling of the dimension arrays d_d? As shown in earlier books, each dimension array has double the rows and double the columns of the array beneath it. Thus each array has four copies of the array beneath it. The origin of the close correspondence between dimension array sizes and Cayley numbers is also an important question.

We provide deeper answers for these questions in this book. *In particular we show Hypercomplex Cosmology is superior to our Octonion Cosmology in several ways including: a more consistent spectrum with respect to space-times, and more importantly, support for Gold Dust[2] as a lower layer of formulation. Octonion Cosmology does not have a similar lower layer.[3]*

FORM OF THE HYPERCOMPLEX SPACES SPECTRUM

Blaha Space Number	Cayley-Dickson Number	Cayley Number	Dimension Array	Space-time-Dimension	CASe Group $su(2^{r/2}, 2^{r/2})$
O_s	n	d_c	d_d	r	CASe
0	10	1024	2048×2048	18	su(512,512)
1	9	512	1024×1024	16	su(256,256)
2	8	256	512×512	14	su(128,128)
3	7	128	256×256	12	su(64,64)
4	6	64	128×128	10	su(32,32)
5	5	32	64×64	8	su(16,16)
6	4	16	32×32	6	su(8,8)
7	**3**	**8**	**16×16**	**4**	**su(4,4)**
8	2	4	8×8	2	su(2,2)
9	1	2	4×4	0	su(1,1)

Figure 1.1. The NEW Hypercomplex Cosmology ten space spectrum. The space for our universe, is number 7, with Cayley-Dickson number 3 (which corresponds to octonions) is in bold type.

---

[2] We use the word "Dust" rather than foam because foam signifies an underlying layer typically of a liquid. Dust does not. There is no underlying layer in Hypercomplex Cosmology.
[3] One cannot chop a spinor array element into pieces in a physically meaningful manner.

2. Unified SuperStandard Theory (NEWUST) and NEWQUeST

Our universe corresponds to Cayley-Dickson number 3 (Blaha number 7) in Fig. 1.1. We use Blaha numbers in addition to Cayley-Dickson numbers because later we show Blaha numbers correspond to the spectrum numbers of a model generating the space spectrum of Fig. 1.1.

Blaha Space 7, which is described in the NEWQUeST theory presented in our books of 20920-2021, was shown to directly correspond to the NEWUST – New Unified SuperStandard Theory – presented in the author's books in the recent past such as Blaha (2018e) and (2020c). We summarize these equivalent theories in Appendix A.

3. Megaverse NEWUTMOST

The NEWUTMOST theory presented in earlier books corresponds to the space of Cayley-Dickson number 4 (Blaha space number 6). It describes a Megaverse (Multiverse). See Appendix B for an outline of details.

4. Enhanced Hypercomplex Cosmology Spectrum

The Hypercomplex Cosmology of Fig. 1.1, which we derived in Blaha (2022), suffices for the description of Cosmology in the large. However, an extension of the spectrum to fractionally dimensioned arrays (d_d) has a deeper layer supporting a new paradigm for the Cosmos.

Its origin is based on the observation that a partial sum of dimension array column (or row) elements (non-negative space-time dimensions) is always less than the number above it by exactly 4 dimensions. For example the sum of the columns in column d_{dc} for Cayley numbers 1 through 5 is 124 while the next column size above it is 128. The complete set of column sizes d_{dc} sums to 4092 while the next column size above it if the spectrum was increased to Cayley number 11 is d_{dc} = 4096. If we embed the product of the 10 spaces in a SU(2048) fundamental representation then there is a shortfall of 2 complex dimensions.

We may remedy the shortfall by extending the spectrum infinitely to fractional dimension arrays as shown in Fig. 4.1. The summation of the additional fractional array columns for space-time dimensions less than zero equals 4 remedying the shortfall:

$$4 = 2 + 1 + \tfrac{1}{4} + 1/8 + 1/16 + \ldots = 2 + 1 + 1$$

The column lengths satisfy the identity

$$\sum_{n=-\infty}^{k} d_{dc}(n) = d_{dc}(k-1) \tag{4.1}$$

where n and k are Blaha space numbers.

The introduction of negative space-time dimensions and of fractional array sizes requires further consideration, which we provide in the following chapters.

EXTENDED FORM OF THE HYPERCOMPLEX SPACES SPECTRUM

Blaha Space Number	Cayley-Dickson Number	Cayley Number	Dimension Array column length	Space-time-Dimension	CASe Group $su(2^{r/2}, 2^{r/2})$
O_s	n	d_c	d_{dc}	r	CASe
0	10	1024	2048	18	su(512,512)
1	9	512	1024	16	su(256,256)
2	8	256	512	14	su(128,128)
3	7	128	256	12	su(64,64)
4	6	64	128	10	su(32,32)
5	5	32	64	8	su(16,16)
6	4	16	32	6	su(8,8)
7	**3**	**8**	**16**	**4**	**su(4,4)**
8	2	4	8	2	su(2,2)
9	1	2	4	0	su(1,1)

EXTENSION:

10	0	1	2	-2	SU(1)?
11	-1	½	1	-4	su(½)?
12	-2	¼	½	-6	•
13	-3	1/8	¼	-8	•
14	-4	1/16	1/8	-10	•

•
•
•

Figure 4.1. The Extended Hypercomplex Cosmology ten space spectrum.

5. Negative Space-time Dimensions

The concept of negative space-time dimensions is new. We define them in terms of a conventional operational definition of dimension.

The dimension of a space-time is the number of independent parameters needed to define a point. An over-determined space-time has a negative dimension.

In particular, we regard a negative space-time dimension as the equivalent of an over-specified point. For example, some negative space-time dimensions for r = -3 through r = 1 are.

r = -3

    x = 3   x = 2*1.5   x = 6/2 x = 12/4   over-determined

r = -2

         x = 3   x = 2*1.5 x = 6/2   over-determined

r = -1

    x = 3   x = 2*1.5   over-determined

r = 0

    x = 3

r = 1 One dimension space with one free parameter

         x   free parameter

where r is the space dimension.

5.1 Other Definitions of Dimension

Other definitions are possible. There are many. We believe the definition above is best suited for Hypercomplex Cosmology because it is in accord with the specification of dimension arrays in Figs. 1.1 and 4.1. The 2×2, 1×1, and fractional array sizes indicate over-determined.

6. Gold Dust

A major benefit of Hypercomplex Cosmology is its support for a deeper level based on the decomposition of particles, operators and dimensions into component parts down to the infinitesimal level. Thus one can view the genesis of Hypercomplex Cosmology as a *local* phenomenon that grows to enormous proportions.

6.1 The Partition of a Unit into Sequence of Space Units

In chapter 4 we described the enhanced Hypercomplex Cosmology spectrum that adds an infinity of spaces to the Hypercomplex Cosmology spectrum. Most of the additional spaces are fractional. Fig. 6.1 depicts the amalgamation of spaces of negative space-time dimension with r < -4 to construct a space-time r = -4 unit space with d_{dc} = 1.

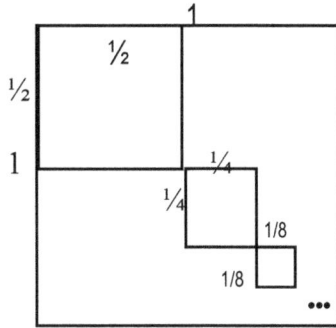

Figure 6.1. Depiction of a amalgamation of an infinite number of spaces for r ≤ - 6 to form a space with a 1×1 dimension array.

The construction of a unit space follows from eq. 4.1:

$$\sum_{n = -\infty}^{12} d_c(n) = d_c(11) = 1 \qquad (6.1)$$

where n, 11 and 12 are Blaha space numbers.

We will use the expansion of the unit space to disassemble fundamental fermions and bosons, raising/lowering operators, and dimensions into Gold Dust.

6.2 Creation of Dust

The creation of Gold Dust follows from repeated use of eqs. 4.1 and 6.1. Consider one item from the sets of fundamental fermions and bosons, raising/lowering operators, and dimensions.

We can use eq. 6.1 to disassemble the item into an infinite set of sub-items as depicted in Fig. 6.1. In chapter 8 we discuss the variety and features of Gold Dust.[4]

Having disassembled an item into the variety exhibited in Fig. 6.1 we can now disassemble the individual sub-items. For example, we can disassemble all sub-items of Blaha space number 11, 12, 13 and 14 into "sub-sub-items" using eq. 4.1. Then all component sub-items are of Blaha number ≥ 15 and extending to ∞. We can call this further subdivision a *level 15 decomposition*.

Clearly we can repeat this decomposition process over and over creating level after level of decomposition. For example the level 17 decomposition of a unit is the sum of all sub-items of Blaha number 17 or below (higher numerically).

Thus every particle, creation operator, annihilation operator, and dimension can be subdivided to dust of various levels without limit. Every item can be viewed as an accretion of dust including the possibility of an infinite accretion of infinitesimal dust "grains."

6.3 Dust Levels

We define dust level d to be the subdivision of an item such that all component items have a $d_{cd} \leq$ d. We say their *fineness* is d.

---

[4] Some types of dust may be non-quantum.

7. Quantum Mechanical Model for Hypercomplex Space Spectrum

7.1   A Bottom Up View of the Spectrum of Spaces

The extended spectrum of spaces of Fig. 4.1 invites the view that it corresponds to an inverted spectrum of states where d_{dc} plays a role analogous to (negative) energy, and the Blaha space number O_s plays the role of an energy level index. Note the energy accumulation point at zero as $O_s \rightarrow \infty$. Note also the geometric scaling behavior of the energy d_{dc}. Fig. 7.1 shows the inverted spectrum plotted as an energy spectrum.

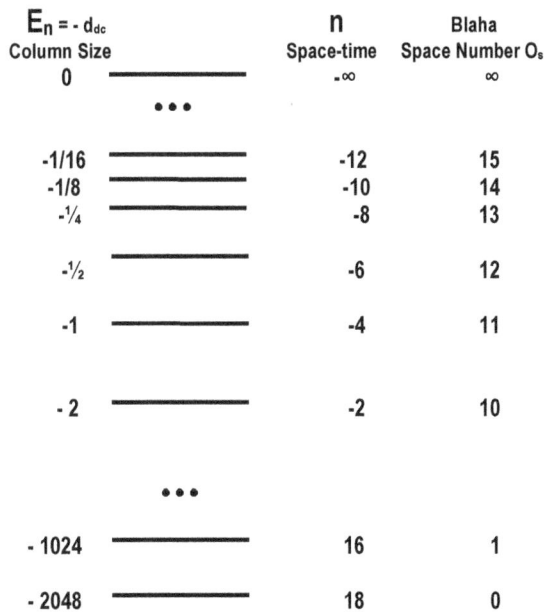

$E_n = -d_{dc}$ Column Size		n Space-time	Blaha Space Number O_s
0	————	$-\infty$	∞
	$\bullet\bullet\bullet$		
-1/16	————	-12	15
-1/8	————	-10	14
$-\frac{1}{4}$	————	-8	13
$-\frac{1}{2}$	————	-6	12
-1	————	-4	11
- 2	————	-2	10
	$\bullet\bullet\bullet$		
- 1024	————	16	1
- 2048	————	18	0

Figure 7.1. Extended Hypercomplex Cosmology spectrum plotted as an Inverted energy spectrum.

7.2 A Model for the Origin of Hypercomplex Spaces

The spectrum in Fig. 7.1 invites a characterization of the Hypercomplex Cosmos spectrum of spaces as resulting from a bound state theory.

We therefore assume that a one-dimension linear "space-time" exists with a potential for a Schrödinger-like equation that provides a bound state spectrum similar to that of Fig. 7.1.

Noting the spectrum of Fig. 7.1 is geometric it is reasonable to consider a Schrödinger equation model with a geometric energy spectrum.[5] The $1/r^2$ model of Case[6] is the only known model. Following Case we find its Schrödinger equation becomes

$$d^2u/dx^2 + (\lambda/x^2 - \eta^2)u = 0 \tag{7.1}$$

where u is the wave function and

$$\eta = r_0(-ME/\hbar^2)^{1/2} = (-E/d)^{1/2} \tag{7.2}$$
$$\lambda' = (\lambda - 1/4)^{1/2} \tag{7.3}$$

and where λ is a dimensionless constant, d is a constant, E is the energy, M is a mass, and r_0 is a coordinate scale. The energy found by Case is

$$\eta_N = \exp[(B - (N + 1/2)\pi)/\lambda'] \tag{7.4}$$

where B is a constant and N is the principal quantum number. Defining

$$N = O_s = \text{the Blaha space number} \tag{7.5}$$
$$\eta_N = d_{dcN}$$

and using

$$N = 0 \quad\quad \eta_N = 2048$$
$$N = 1 \quad\quad \eta_N = 1024$$

to set parameters in eq. 7.4 we find

$$\lambda' = 4.532 = \pi/\ln(2) \tag{7.6}$$
$$\lambda = 20.792$$
$$\eta_N = 2^{(11-N)}$$

and

$$E_N = -d\eta_N^2 = -d2^{(22-2N)} \tag{7.7}$$

Using

$$O_s = \tfrac{1}{2}(18 - r) \tag{7.8}$$

as is evident in Fig. 4.1, and setting d = 1 we find

$$E_N = -\eta_N^2 = -2^{r+4} \tag{7.9}$$

$$d_{dcN} = 2^{11 - O_s} \tag{7.10}$$

---

[5] A quark spectrum with geometric scaling behavior was considered in the author's papers: "Is Flavor Fundamental" Phys. Lett. B **80**, 99 (1978) and "On a Possible Similarity Between the Heavy Lepton and Heavy Constituent Quark Mass Spectra" Phys. Lett. B **84**, 116 (1979)
[6] K. M. Case, Phys Rev, **80**, 797 (1950).

$$d_{dcN} = 2^{r/2 + 2} \tag{7.11}$$

From eq. 7.9 we see E_{O_s} is the negative of the O_s space dimension array size in Fig. 4.1 if we treat the dimension array of the space as a long linear array of dimensions. (Normally the array is viewed as a square array.) We conclude:

1. The energy spectrum sets the size of the spectrum of space dimension arrays if viewed as long linear arrays.
2. The upper bound on the space spectrum of $O_s = 0$ giving a 10 space spectrum neglecting the extension is set by the lowest energy level.
3. The extension is physically understandable as part of a more general view based on a scaling $1/r^2$ potential.
4. The space of the Schrödinger is a one-dimension line. The energy spectrum it generates is a one dimension array of dimension numbers. A consistency.

This model displays the extreme localization of the $1/r^2$ potential and the consequent self-similar, scaling behavior.

7.3 Scale Breaking

The scaling potential reflects the lack of scale in the one-dimension space. Eq. 7.1 reveals the scale invariance of the model. Scale invariance requires the energy levels index extend from $N = -\infty$ to $N = +\infty$. The energy levels in eq. 7.4 are invariant under shifts of the constant B by $k\pi$ where k is a positive or negative integer—discrete scaling.

The model reflects the scale invariance. Consider a change of scale in x by letting

$$x \rightarrow e^{k\pi/\lambda'}x \tag{7.12}$$

where k is an integer. Then

$$\eta \rightarrow e^{k\pi/\lambda'}\eta \tag{7.13}$$

in eq. 7.1 and the energy levels remain the same by eqs. 7.2 and 7.4 requiring only a superficial shift in N.

In order to obtain the energy spectrum of Fig. 7.1 we must break scale invariance by restricting N to non-negative integers in eqs. 7.7 and 7.9, thus preventing shifts in the B constant by scaling as in eq. 7.12.

Thus the restriction of the Hypercomplex Cosmology spectrum to 10 spaces reflects a breakdown in scale invariance in this model.

7.4 Removing the Negative Index N Values

The values of $N < 0$ can be removed from the spectrum by supplementing Case's requirement of a fixed phase for wave functions at the origin with the requirement of the Bohr angular momentum quantization condition: $p_\varphi = mvr = N\hbar$ where N is non-negative. (Eq. 7.1 originated in a 3D Schrödinger equation as noted earlier.) Thus the energy levels of Fig. 7.1 result. (Note that the level numbering could be reset to start at $N = 1$ using a shift in the value of B.)

7.5 Cosmos Construction

The picture portrayed in this chapter suggests a Cosmos originating (without time evolution) from a linear one-dimension space with a scaling potential that "confines" the set of spaces with positive space-time dimensions to 10 spaces. The number of space-time dimensions in a space is given by eq. 7.8. The total number of dimensions in the dimension array of a space is equal to the energy in the above model. Its geometric nature follows from the $1/r^2$ potential, which in turn reflects a scaling property of the one-dimension model.

For reasons given in Blaha (2022) the space-time dimensions must be even numbers. They start from a 0 dimension space-time and go to 18.

After determining the dimension arrays of the spaces, the internal symmetry content of each space is determined by self-similarity with each space having a fourfold set of the space below it in Fig. 1.1. The number of rows and columns of each dimension array increases by a factor of two as one ascends the spaces listed in Fig. 1.1.

Blaha (2022) derived the set of spaces in Fig. 1.1 by considering the creation/annihilation operators a fermion in each space. The CASe group column in Fig. 1.1 shows the group obtained from the consideration of CASe transformations on the creation/annihilation operators of a fermion due to General Relativistic transformations.

The equivalence of the General Relativistic derivation of the spaces spectrum of Fig. 1.1 and the Schrödinger model derivation above is encouraging. Both are based on self-similarity. We thus see Hypercomplex Cosmology as a consequence of scaling and self-similarity in a precursor "space."

The creation of universes of a space type containing matter and energy follows directly as we showed previously using fermion-antifermion annihilation. A universe may be created in a Fundamental Frame (Blaha (2022)) that becomes an evolving universe in a static frame that is populated with multiple copies of the Fundamental Frame particles and symmetries as we saw in Blaha (2022). Thence expansion and evolution.

Hypercomplex Cosmology provides a complete scenario for the Cosmos.

7.6 Critical Role of Creation/Annihilation Operators

The connection of the model presented above and the creation/annihilation basis of Hypercomplex Cosmology is self-similarity and scaling.

The creation/annihilation operator basis is of primary importance. Numerically for each space, the number of fermion creation/annihilation operators equals the number of array dimensions, which equals the number of fundamental fermions, which in turn equals the number of vector bosons.

In addition we show in chapter 10 that one can establish a computer paradigm for each space with computer languages based on its creation/annihilation operator expressions.

8. Gold Dust Features

8.1 Types of Gold Dust

Earlier in chapter 6 we showed how to fractionate gold dust into various levels of fineness. The process could be applied to individual or sets of creation/annihilation operators, fermions, bosons, and dimensions.

The types of the quantities subject to fractionalization are:

> Creation operators of all internal symmetries and spins
> Annihilation operators of all internal symmetries and spins
> Fundamental fermion operators of all internal symmetries and spins
> Fundamental boson operators of all internal symmetries and spins
> Dimensions

Gold dust of the first four types is quantum with commutation rules. Dimension gold dust is not quantum.

The fractionation of a particle reduces a particle to an aggregate of "dust" where each infinitesimal part contains part of the particle's charge and internal symmetries.

8.2 Confinement of Gold Dust

A fundamental particle, as we see it, is a unitary entity that does not appear to be composite although there are theories that subdivide particles. We are somewhat familiar with quark confinement, a related phenomenon, although the confinement mechanism is problematic. Some like Kenneth Wilson (Cornell and Ohio State) suggest a lattice approach. Some, like the author, suggest it is dynamic through higher derivative strong interactions. (Wilson has suggested to the author that the author's theory is a phenomenology for his lattice theory. *C'est la vie*!)

We now have a new confinement problem: How do particles, which are subject to fractionation as we suggested earlier, cohere as unitary objects to become fundamental particles of NEWQUeST (NEWUST) and NEWUTMOST?

A particle, composed of Gold Dust, presumably has total dust confinement. The dust may be confined by a force or by being enclosed in an impenetrable membrane or by confinement through some lattice theory.

Gold Dust confinement thus raises an important issue. Perhaps it is an issue that experiment has raised already but we were unable to recognize it. Perhaps parton models of Deep Inelastic Scattering were about fractionated quarks. It is known that parton models have some issues when matched to experiment. Perhaps the quarks in a proton or neutron are being viewed at higher precision (higher fineness) revealing their internal fractional gold dust components. See section 6.3.

Gold dust "glue" must be analogous to color confinement "glue." Then confined Gold Dust aggregates to form particles.

8.3 The Profile of a Particle as it Proceeds to Higher Fineness through Increased Fractionation

If we fractionate a fundamental particle we may expect the particle to pass through a number of stages if we could view it in a microscope of increasing resolution. First it becomes a set of quantum "sub-particles. Then it passes through the stages:

> Quantum liquid – with a lumpiness still
> Classical liquid – totally a smooth "classical" fluid

In the final states of fractionation the Gold Dust is so fine that it appears to be a classical fluid thus accomplishing the reduction of a particle to a fluid of infinitesimal parts. The fluid (dust) has spins, internal quant numbers, species types, and energies distributed throughout.

The Gold Dust combines to produce the creation/annihilation operators of particles. We viewed the particles of a space as created in a Fundamental Frame in Blaha (2029) at the origin point of a universe. (Universes begin as a point composed of dust.) At the origin of a space (universe) we see a spontaneous condensation of dust to particles. Then splitting of the space into copies occurs as described in Blaha (2022) followed by dynamical evolution with interactions as the universe expands.

8.4 Commutation Rules for Particle Gold Dust

At every level of fineness we can define fractional particle states and creation/annihilation operator commutation rules. To create a particle fraction from the vacuum state $|0>$ we define a $1/n$ fraction state

$$|1/n> = b^{1/n\dagger}|0> \tag{8.1}$$

and we define its annihilation with

$$b^{1/n}|1/n> = |0> \tag{8.2}$$

Then, for multiple fraction states of fractions $1/n$ and $1/m$, we see on physical grounds:

$$(b^{1/m})^j (b^{1/n\dagger})^k|0> \propto (b^{k/n - j/m})^\dagger |0> \tag{8.3}$$

if $k/n - j/m \geq 0$. If $k/n - j/m < 0$, then

$$(b^{1/m})^j (b^{1/n\dagger})^k|0> \propto b^{j/m - k/n} |0> = 0 \tag{8.4}$$

If $k/n - j/m = 0$ then $(b^{1/m})^j (b^{1/n\dagger})^k|0> = (\text{const}) |0>$.

With this basis one can define commutation (or anticommutation) relations for fractional operator.

9. A New Level for Hypercomplex Cosmology

All creation is dust.

The discovery of the basis of Hypercomplex Cosmology features in Gold Dust implies a new foundation for Cosmology in an infinitesimally grained substratum. This substratum is very credible from a Philosophic viewpoint since it avoids instant creation by *fiat*. The Cosmos grows infinitesimally—space by space, dimension by dimension, fractionally by creation operators, fractionally by particles, and fractionally in interactions. Annihilation takes place infinitesimally. The Cosmos is generated *locally in a manner appealing to the continuing local trend of Physics.*

The infinitesimal nature of creation and annihilation is masked by forms of aggregated confinement enforced by dynamics.

Gold Dust opens the issues of the binding of particles and the nature of negative space-time dimensions. Importantly, it raises the question of the confinement of Gold Dust within spaces, dimensions, and, most importantly, in fermion and boson particles.

The central role of creation and annihilation operators also leads to the consideration of such operators for the creation and annihilation of spaces (universes), which we considered in Blaha (2018e).

Our development of Hypercomplex Cosmology leads to a view of the Cosmos as composed of spaces (universes) that are dust, of dimensions that are dust, of universes of galaxies, clouds, stars and planets that are dust, of matter and energy that are dust, and of interactions that are dust. All things are Gold Dust united according to confinement and dynamics at multiple levels.

Looking backwards from the spaces (universes) of Hypercomplex Cosmology we see that their ultimate origin in Gold Dust provides a graduated platform of growth from "Nothingness."

10. Computer Form of the Cosmos

Once particles have their complete form as particles and not as accretions of dust, we can develop a computation model of particles and their interactions in each Hypercomplex Cosmology space (universe).

This chapter describes the creation of computer programs that reproduce the dynamical features of Hypercomplex Cosmology.

It begins by identifying the "memory" of a Hypercomplex space. Then it defines computation operators corresponding to features of Hypercomplex Cosmology. Almost all of the passages below appear in Blaha (1998) and (2005c). We frame the discussion in terms of a Quantum Computer.

Each space (universe) of Hypercomplex Cosmology has fermion and boson particles defined including fermions, Higgs bosons, and the vector bosons of interactions and gravity. The number of types of each category of particle is finite.

For example there are 256 different fermions in our universe in NEWQUeST. The set of fermions defines a memory. In our universe the fermion memory has 256 slots. Each set of bosons also has memory. Memories can be populated with particles at space-time points.

Then we can view particle interactions that create and annihilate particles as memory manipulations. Similarly, as we see below, we can formulate computations using raising and lowering operators that are analogous to the creation/annihilation operators that formed the foundation of Hypercomplex Cosmology. Computer memory[7] represents the vacuum!

The comparison of creation/annihilation operators with raising/lowering operators is deepened by considering the process of setting a data value in computer memory. Setting a data value costs energy just as creating a particle costs energy. Erasing a data value also costs energy as does annihilating a particle.

Recent experiments have shown that a logical value of a qubit data value has an energy associated with it. One bit of information has about 3×10^{-21} joules of energy[8] or a rest mass, m_0, or about 0.02 eV using $m_0 = E/c^2$. This result was confirmed by E. Lutz et al.[9] who showed that there is a minimum amount of heat produced per bit of erased data. This minimal heat is called the *Landauer[10] limit*.

---

[7] The dimension array for each particle type of each space is a computer memory specification for each point of the space.

[8] E. Muneyuki et al, *Nature Physics*, DOI: 10.1038/NPHYS1821.

[9] E. Lutz et al, Nature **483** (7388): 187–190,10.1038/nature10872, (2012).

[10] R. Landauer, "Irreversibility and heat generation in the computing process", IBM Journal of Research and Development **5** (3): 183–191, (1961).

Thus the analogue between particle creation/annihilation and data creation/annihilation via computer instructions or as implemented below with creation/annihilation operators is firm.

We see a Computation Model in the future of Hypercomplex Cosmology. This viewpoint leads to a view of a space as a type of computer with memories and programs managing memories.

We now develop Quantum Computation from the Hypercomplex Cosmology perspective. We find creation/annihilation operator expressions that are equivalent to standard mathematical and logic operators (instructions) in computer languages.[11]

10.1 Introduction

A natural question that arises when one considers Quantum Computers is the role of the Quantum Computer processor and the operations it supports. A further question of some interest is whether a quantum machine language exists and what its nature might be. Lastly the question of higher level languages is also relevant. Can we develop a Quantum Assembly Language? What is the nature of High Level Quantum Languages? Are there, for example, equivalents to the C or C++ languages?

We develop answers to these questions within the framework of Hypercomplex Cosmology.

10.2 Computer Machine and Assembly Languages

A traditional (non-quantum) computer may be viewed simply as a main memory, an accumulator or register (modern computers have many registers), and a central processing unit (CPU) that executes a program (instructions) step by step. It can be visualized as in Fig. 10.1.

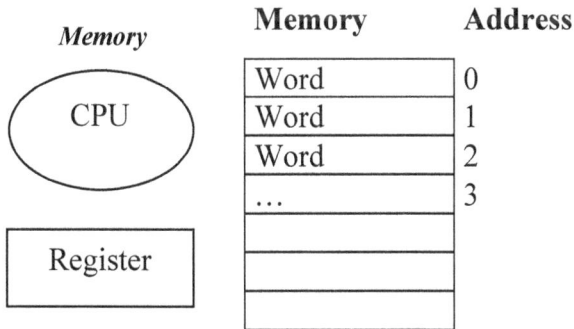

Figure 10.1. A simplified model of a normal computer.

A set of data and a program (or set of instructions) is stored in memory and the CPU executes the program step by step using the data to produce an output set of data.

The basic instructions of assembly language and machine language move data values between memory and the register (or registers), manipulate the data value in the register and provide basic arithmetic and logical operations[12]:

---

[11] The remainder of this Chapter is extracted from the author's 1998 book: *Cosmos and Consciousness.*

LOAD M –   load the value at memory location M into the register

STORE M –   store the value in the register at memory location M

SHIFT k –   shift the value in the register by k bits

The following arithmetic instructions modify the value in the register. The AND, OR and NOT instructions perform bit-wise and, or and not operations.

ADD M   – add the value at memory location M to the value in the register

SUBTRACT M – subtract the value at memory location M from the value in the register

MULTIPLY M – multiply the value in the register by the value at memory location M

DIVIDE M – divide the value in the register by the value at memory location M

AND M   –   change the value in the register by "anding" it with the value at memory location M

OR M –   change the value in the register by "oring" it with the value at memory location M

NOT –   change the value in the register by "not-ing" it

The following instructions implement input and output of data values.

INPUT M –   input a value storing it at memory location M

OUTPUT M – output the value at memory location M

A computer has another register (memory location) called the Program Counter. The value in the program counter is the memory location of the next instruction to execute. The following instructions support non-sequential flow of control in a program. A program can "leap" from one instruction in a program to another instruction many steps away and resume normal sequential execution of instructions.

---

[12] See for example Kurt Maly and Allen R. Hanson, *Fundamentals of the Computing Sciences* (Prentice-Hall, Inc., Englewood Cliffs, NJ, 1978) Chap. 8.

TRA M   – set the value of the program counter to the value at memory location M

TZR M   – set the value of the program counter to the value at memory location M if the value in the register is zero.

HALT   – stop execution of the program

The above set of instructions constitutes an extremely simple assembly language. They also are in a one-to-one correspondence with machine instructions (machine language). Most assembly and machine languages have a much more extensive set of instructions.

10.3 Algebraic Representation of Assembly Languages

The normal view of assembly language is that it has a word or instruction oriented format. Some assembly language programmers would even say that assembly language is somewhat English-like in part.

Computer languages in general have tended to become more English-like in recent years in an attempt to make them easier for programmers. Some view a form of highly structured English to be a goal for computer programming languages.

In this section we follow the opposite course and show that computer languages can be reduced to an algebraic representation. By algebraic we mean that the computer language can be represented with operator expressions using operators that have an algebra similar to that of the raising and lowering (creation/annihilation) operators seen earlier. We will develop the algebraic representation for the case of the simple assembly language described in the previous section. There are a number of reasons why this reduction is interesting:

1. It may help to understand Hypercomplex Cosmology dynamics more deeply (later in this chapter).

2. It will deepen our understanding of computer languages.

3. It provides a basis for the understanding of Quantum Computers.

4. It may have a role in research on one of the major questions of computer science: proving a program actually does what it is designed to do. Algebraic formalisms are generally easier to prove theorems then English-like formalisms.

The algebraic representation can be defined at the level of individual bits based on anti-commuting Fermi operators. But it seems more appropriate to develop a representation for "words" consisting of some number of bits. An algebraic representation for a word-based assembly language can be developed using commuting harmonic oscillator-like raising and lowering operators.

A word consists of a number of bits. In currently popular computers the word size is 32 bits (32-bit computer). The size of the word determines the largest and smallest integer that can be stored in the word. The largest integer that can be stored in a 32-bit word is 4,294,967,294 and the smallest integer that can be stored in a 32-bit word is 0 if we treat words as holding unsigned integers.

To develop a simple algebraic representation of assembly language we will assume the size of a word is so large that it can be viewed as infinite to a good approximation. (It is also possible to develop algebraic representations for finite word sizes.) As a result memory locations can contain non-negative integers of arbitrarily large value.

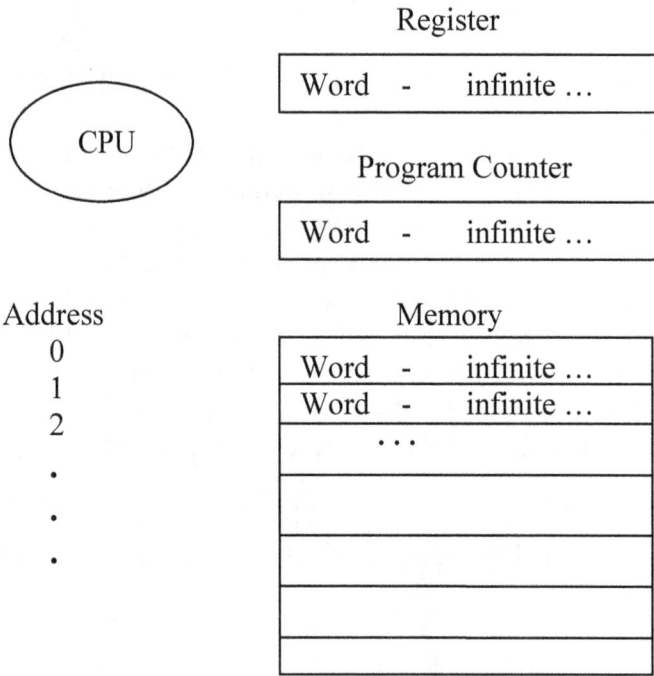

Figure 10.2. Visualization of a Computer with infinite words.

To establish the algebraic representation we associate a harmonic raising operator $a_i^\dagger$ and a lowering operator a_i with each memory location. These operators satisfy the commutation relations:

$$[\, a_i, a_j^\dagger\,] = \delta_{ij}$$
$$[\, a_i, a_j\,] = 0$$
$$[\, a_i^\dagger, a_j^\dagger\,] = 0$$

where δ_{ij} is 1 if i = j and zero otherwise. We define a pair of raising and lowering operators for the register r and $r^\dagger$ with commutation relations

$$[\, r_i, r_j^\dagger \,] = \delta_{ij}$$
$$[\, r_i, r_j \,] = 0$$
$$[\, r_i^\dagger, r_j^\dagger \,] = 0$$

The ground state of the computer is the state with the values at all memory locations set to zero. It is represented by the vector

$$|\, 0, 0, 0, \ldots > \equiv |\, 0 > \equiv \Phi_V$$

A state of the computer will be represented by a vector of the form

$$|\, n, m, p, \ldots > \; = \; N \, (r^\dagger)^n (a_0^\dagger)^m \, (a_1^\dagger)^p \ldots |\, 0 >$$

where N is a normalization constant and with the first number being the value in the register, the second number the value at memory location 0, the third number the value at memory location 1, and so on. For simplicity we will not consider superpositions of computer states at this point. We will discuss superpositions later in this chapter. Within this limitation we can set a computer state to have certain initial values in memory and then have it evolve by executing a "program" to a final computer state with a different set of computer values in memory. The "program" is a mapping of the instructions of an assembly language program to algebraic expressions in the raising and lowering operators.

10.4 Basic Operators of the Algebraic Representation
The key operators that are required for the algebraic representation are:

Fetch the Value at a Memory Location (Number Operator)

$$N_m = a_m^\dagger a_m$$

For example,

$$N_m |\, \ldots, n, \ldots > \; = n |\, \ldots, n, \ldots >$$

$\uparrow$

m^{th} memory location value

Set the Value at Memory Location m to Zero

$$M_m = \frac{(a_m)^{N_m}}{\sqrt{N_m!}}$$

The above expression for M_m is symbolic. The expression represents the following expression in which the operators are carefully ordered to avoid complications (c-numbers etc.) resulting from reordering.

$$M_m \equiv \sum_q \frac{(\ln a_m)^q N_m^{\ q}}{q!} \frac{1}{\sqrt{N_m!}}$$

where the sum ranges from 0 to ∞. When M_m is applied to a state it sets the value of the mth memory location to zero.

$$M_m | \ldots, n, \ldots > = \frac{(a_m)^n}{\sqrt{n!}} | \ldots, n, \ldots >$$

$$\uparrow$$

$$m^{th} \text{ memory location value}$$

$$= | \ldots, 0, \ldots >$$

The repeated application of factors of N_m to the state results in factors of n.

Change Value at Memory Location m from 0 to Value at Location n

$$P_m^{\ n} = \frac{(a_m^\dagger)^{N_n}}{\sqrt{N_n!}}$$

The above expression for $P_m^{\ n}$ is also symbolic. The expression represents the following expression in which the operators are carefully ordered to avoid complications (c-numbers etc.) resulting from reordering.

$$P_m^{\ n} \equiv \sum_q \frac{(\ln a_m^\dagger)^q N_n^{\ q}}{q!} \frac{1}{\sqrt{N_n!}}$$

where the sum ranges from 0 to ∞. When $P_m^{\ n}$ is applied to a state it changes the value of the mth memory location from zero to the value at the nth memory location.

$$m^{th} \qquad n^{th}$$

$$P_m^{\,n} \,|\, \ldots, 0, \ldots, x, \ldots \,> \;=\; \frac{(a_m^{\dagger})^x}{\sqrt{x!}} \,|\, \ldots, 0, \ldots, x, \ldots >$$

$$= \; |\, \ldots, x, \ldots, x, \ldots >$$

The application of the factors of N_n to the state results in factors of x that lead to the above expression when summed.

The operators M_m and $P_m^{\,n}$ enable us to simply express the algebraic equivalent of assembly language instructions:

LOAD m – load the value at memory location m into the register

$$P_r^{\,m} \, M_r$$

STORE m – store the value in the register at memory location m

$$P_m^{\,r} \, M_m$$

SHIFT k – shift the value in the register by k bits. If k is positive the bit shift is to the right and if k is negative the bit shift is to the left. The bits are numbered from the leftmost bit which is bit 0 corresponding to 2^0. The next bit is bit 1 corresponding to 2^1 and so on.

If the bit shift is to the right (k > 0) then we assume the padding bits are 0's. For example a shift of the bit pattern for 7 = 1110000 … one bit to the right is 14 = 01110000 … As a result the value in the register is doubled (k = 1), quadrupled (k = 2), and so on. The algebraic expression for a k bit right shift is

$$\sum_q \frac{(\ln a_r^{\dagger})^q \, S_r^{\,q}}{q!} \; \frac{\sqrt{N_r!}}{\sqrt{T_r!}}$$

where

$$S_r = (2^k - 1)N_r$$

and

$$T_r = 2^k N_r$$

If the bit shift is to the left (negative k), then we assume zero bits are added "at ∞". If k = -1 then the effect of left shift is to divide the value in the register by two (dropping the fractional part). If k = -2 then the effect of left shift is to divide the value in the register by four (dropping the fractional part) and so on. The algebraic expression that implements left shift is

$$\sum_q \frac{(\ln a_r)^q \, U_r^{\,q}}{q!} \, \frac{\sqrt{N_r!}}{\sqrt{V_r!}}$$

where

$$U_r = N_r - [\, 2^k N_r \,]$$

and

$$V_r = [\, 2^k N_r \,]$$

with [z] being the value of z truncated to an integer (fractional part dropped).

ADD m   –   add the value at memory location m to the value in the register

$$\sum_q \frac{(\ln a_r^\dagger)^q \, N_m^{\,q}}{q!} \, \frac{\sqrt{N_r!}}{\sqrt{(N_r + N_m)!}}$$

SUBTRACT m   – subtract the value at memory location m from the value in the register (assumes the value in the register is greater than or equal to the value at location m)

$$\sum_q \frac{(\ln a_r)^q \, N_m^{\,q}}{q!} \, \frac{\sqrt{N_r!}}{\sqrt{(N_r - N_m)!}}$$

MULTIPLY m   – multiply the value in the register by the value at memory location m

$$\sum_q \frac{(\ln a_r^\dagger)^q \, (N_r)^q (N_m - 1)^q \, \sqrt{N_r!}}{q!} \, \frac{1}{\sqrt{(N_r N_m)!}}$$

DIVIDE m   – divide the value in the register by the value at memory location m

$$\sum_q \frac{(\ln a_r^\dagger)^q \, W^{\,q}}{q!} \, \frac{\sqrt{N_r!}}{\sqrt{X!}}$$

where

$$W = N_r - [\, N_r / N_m \,]$$

and

$$X = [\, N_r / N_m \,]$$

with [z] being the value of z truncated to an integer (fractional part dropped).

AND m – change the value in the register by "and-ing" it with the value at memory location m

$$\sum_q \frac{((\ln a_r^\dagger)^q (W)^q \theta\, (W) + (\ln a_r)^q\, (-W)^q \theta\, (-W)\,)}{q!} \frac{\sqrt{N_r!}}{\sqrt{X!}}$$

where

$$W = N_r \, \& \, N_m - N_r$$

and where

$$X = N_r \, \& \, N_m$$

with $\theta(z) = 1$ if $z > 0$ and 0 if $z < 0$. The & operator (adopted from the C programming language) performs bitwise AND. Corresponding bits in each operand are "multiplied" together using the multiplication rules:

$$1 \, \& \, 1 = 1$$
$$1 \, \& \, 0 = 0 \, \& \, 1 = 0 \, \& \, 0 = 0$$

For example the binary numbers 1010 & 1100 = 1000 or in base 10 5 & 3 = 1.

OR m – change the value in the register by "or-ing" it with the value at memory location m

$$\sum_q \frac{(\ln a_r^\dagger)^q (W)^q}{q!} \frac{\sqrt{N_r!}}{\sqrt{X!}}$$

where

$$W = N_r \,|\, N_m - N_r$$

and where

$$X = N_r \,|\, N_m$$

The | operator (adopted from the C programming language) performs bitwise OR. Corresponding bits in each operand are "multiplied" together using the multiplication rules:

$$1 \mid 1 = 1 \mid 0 = 0 \mid 1 = 1$$
$$0 \mid 0 = 0$$

For example the binary numbers $1010 \mid 1100 = 1110$ or in base 10 $5 \mid 3 = 7$.

NOT – change the value in the register by "noting" it

$$\sum_q \frac{((\ln a_r^\dagger)^q (W)^q \theta(W) + (\ln a_r)^q (-W)^q \theta(-W))}{q!} \frac{\sqrt{N_r!}}{\sqrt{X!}}$$

where

$$W = \sim N_r - N_r$$

and where

$$X = \sim N_r$$

with $\theta(z) = 1$ if $z > 0$ and 0 if $z < 0$. The $\sim$ operator (adopted from the C programming language) performs bitwise NOT. Each 1 bit is replaced by a 0 bit and each 0 bit is replaced by a 1 bit. Since we have infinite words in our computer we supplement this rule with the restriction that the exchange of 1's and 0's only is made up to and including the rightmost 1 bit in the operand. The 0 bits beyond that remain 0 bits. For example the binary number $\sim 101 = 010$ or in base 10, $\sim 3 = 2$.

INPUT m – input a value storing it at memory location m. The input device is usually associated with a memory location from which the input symbolically takes place. We will designate the memory location of the input device as *in*.

$$P_m^{\ in} M_m$$

OUTPUT m – output the value at memory location m. The output device is usually associated with a memory location to which output symbolically takes place. We will designate the memory location of the output device as *out*.

$$P_{out}^{\ m} M_{out}$$

TRA m – set the value of the program counter to the value at memory location m. If we designate the program counter memory location as pc then this instruction is mapped to

$$P_{pc}^{\ m} M_{pc}$$

TZR m –   set the value of the program counter to the value at memory location m if the value in the register is zero.

$$(P_{pc}{}^m \; M_{pc})^{\; \theta(N_r) \; \theta(-N_r)}$$

using $\theta(0) = 1$.

HALT –   stop execution of the program. The halt in a program is mapped to a "bra" state vector.

$$< \dots \mid$$

10.4.1 A Simple Assembly Language Program

Assembly language instructions can be combined to form an assembly language program. Perhaps the best way to see how the algebraic representation of assembly language works is to translate a simple assembly language program into its algebraic equivalent.

The program that we will consider is:

1	INPUT x
2	INPUT y
3	LOAD x
4	ADD y
5	STORE z
6	OUTPUT z
7	HALT

This program translates to the algebraic equivalent:

$$\underset{7}{} \quad \underset{6}{} \quad \underset{5}{} \quad \underset{4}{} \quad \underset{3}{} \quad \underset{2}{} \quad \underset{1}{} \quad \text{Steps}$$

$$< \dots \mid P_{out}{}^z \, M_{out} \quad P_z{}^r \, M_z \quad \frac{(a_r{}^\dagger)^{N_y} \sqrt{N_r!}}{\sqrt{(N_r + N_y)!}} \quad P_r{}^x \, M_r \quad P_y{}^{in} \, M_y \quad P_x{}^{in} \, M_x \mid \dots >$$

where the power of $a_r{}^\dagger$ is represented by a power series expansion as seen earlier.

The algebraic expression in the brackets produces one output state from a given initial state. The values in memory after the last step correspond to one and only one output state of the form:

$$< n, m, p, \; \dots \mid \; = \; (N \, (r^\dagger)^n (a_0{}^\dagger)^m \, (a_1{}^\dagger)^p \dots \mid 0 >)^\dagger$$

where N is a normalization constant.

This simple program does not produce a superposition of states. As a result programs of this type are analogous to ordinary programs for normal, non-Quantum computers. The numbers in memory after the program concludes are the "output" of the program. We will see programs in succeeding sections that take a computer of fixed state $N\ (r^\dagger)^n (a_0^\dagger)^m\ (a_1^\dagger)^p\ \ldots\ |\ 0 >$ and produce a superposition of states that must be interpreted quantum mechanically. These programs are quantum in nature and the computer that runs them must be a quantum computer.

10.4.2 Programs and Program Logic

The simple program of the last section corresponded to a sequential program that executed step by step. We now turn to more complex programs with program logic that supports non-sequential execution of programs. When this type of program executes the execution of the instructions can lead to jumps from one instruction to another instruction in another part of the program.

Programs are linear – one instruction executes after another. But they are not sequential – the instructions do not always execute step by step sequentially. A program can specify jumps ("goto" instructions) in the code from the current instruction to an instruction several steps after the current instruction or several steps back to a previous instruction. The code then executes sequentially until the next jump is encountered.

These jumps in the code at the level of assembly language implement the control constructs such as goto statements, if expressions, for loops, and switch expressions seen in higher level languages such as C and C++.

Jumps in code can be implemented in the algebraic representation of programs by having a program counter memory value that increments as the algebraic factor corresponding to each step executes. Steps in the program can execute or not execute depending on the current value of the program counter.

Changes in the program counter value are made using the TRA and TZR instructions. In the algebraic representation the program counter variable can be used to manage the execution of the program steps.

The key algebraic constructs supporting non-sequential program execution are:

Execute instruction only if PC $\leq$ n

$$(\ldots)^{\theta\,(n\, -\, N_{pc})}$$

Execute instruction only if PC $\geq$ n

$$(\ldots)^{\theta\,(N_{pc}\, -\, n)}$$

Execute instruction only if PC $=$ n

$$(\ldots)^{\theta\theta\,(N_{pc}\, -\, n)}$$

Execute instruction only if PC not equal to n

$$(\ldots)^{\theta(N_{pc}\, -\, n)\, +\, \theta(n\, -\, N_{pc})\, -\, 2\theta\theta(N_{pc}\, -\, n)}$$

where the parentheses contain one or more instructions and where $\theta\theta(x) = 1$ if $x = 0$ and zero otherwise. The function $\theta\theta(x)$ can be represented by step functions as

$$\theta\theta(x) = \theta(x)\, \theta(-x)$$

Using these constructs we can construct non-sequential programs that support "goto's", if's and other control constructs seen in higher level languages.

To illustrate this feature of the algebraic representation we will consider an enhancement of the assembly language program seen earlier:

```
1          INPUT x
2          INPUT y
3          LOAD x
4          TZR y
5          ADD y
6          STORE z
7          OUTPUT z
8          HALT
```

This program has the new feature that if the first input – to memory location x – is zero, then instruction 4 will cause a jump to the instruction specified by the value stored at memory location y.

For example if the inputs are 0 placed at memory location x and 2 placed at memory location y, then the TZR instruction will cause the program to jump to instruction 2 from instruction 4. Then the program will proceed to execute from instruction 2.

Another example of a case with a jump is if the input to memory location x is zero and the input to memory location y is 6 then the program jumps from instruction 4 to instruction 6 and the program completes execution from there. If the input to memory location x is non-zero no jump takes place.

To establish the algebraic equivalent of the preceding example we have to use the non-sequential constructs provided earlier in this section. In addition we must define the equivalent recursively because of the possibility that the program may jump backwards to an earlier instruction in the program. If only "forward" jumps were allowed then recursion would not be needed.

An algebraic representation of the program that supports only forward leaps is:

$$
\overset{8}{<} \ldots \mid \overset{7}{(a_{pc}^{\dagger} P_{out}^{z} M_{out})}{}^{\theta\theta(N_{pc}-7)}
$$

$$
\overset{6}{(a_{pc}^{\dagger} P_z^{r} M_z)}{}^{\theta\theta(N_{pc}-6)}
$$

$$(\frac{a_{pc}^{\dagger} (a_r^{\dagger})^{N_y} \sqrt{N_r!}}{\sqrt{(N_r + N_y)!}})^{\overset{5}{\theta\theta(Npc-5)}}$$

$$(a_{pc}^{\dagger})^{1-\theta\theta(N_r)} \overset{4}{(P_{pc}^{y} M_{pc})}^{\theta\theta(N_r)\, \theta\theta(Npc-4)}$$

$$\overset{3}{a_{pc}^{\dagger} P_r^{x} M_r}$$

$$\overset{2}{a_{pc}^{\dagger} P_y^{in} M_y}$$

$$\overset{1}{a_{pc}^{\dagger} P_x^{in} M_x}$$

$$a_{pc}^{\dagger} M_{pc} \mid \ldots >$$

The program steps are numbered above each corresponding expression. The step function expressions enable the jump to take place successfully.

A program with forward and backward jumps supported requires a recursive definition. We will define the recursive function f() with:

$$f() = \overset{7}{(a_{pc}^{\dagger} P_{out}^{z} M_{out})}^{\theta\theta(Npc-7)}$$

$$\overset{6}{(a_{pc}^{\dagger} P_z^{r} M_z)}^{\theta\theta(Npc-6)}$$

$$(\frac{a_{pc}^{\dagger} (a_r^{\dagger})^{N_y} \sqrt{N_r!}}{\sqrt{(N_r + N_y)!}})^{\overset{5}{\theta\theta(Npc-5)}}$$

$$(a_{pc}^{\dagger})^{1-\theta\theta(N_r)} \overset{4}{(f() P_{pc}^{y} M_{pc})}^{\theta\theta(N_r)\, \theta\theta(Npc-4)}$$

$$\overset{3}{(a_{pc}^{\dagger} P_r^{x} M_r)}^{\theta\theta(Npc-3)}$$

$$2$$
$$(a_{pc}^{\dagger} P_y^{in} M_y)^{\theta\theta(Npc^- 2)}$$

$$1$$
$$(a_{pc}^{\dagger} P_x^{in} M_x)^{\theta\theta(Npc^- 1)}$$

The program is

$$f()a_{pc}^{\dagger} M_{pc} | \ldots >$$

This program is well behaved except if the input value placed at the y memory location is 4. In this case the program recursively executes forever. This defect can be removed by using another memory location for a counter variable.

We can modify the program so that the program only recursively calls itself a finite number of times by having each recursive call decrease the counter variable by one. When the value reaches zero the recursion terminates. An example of such a program (set to allow at most 10 iterations of the recursion) is:

$$7$$
$$g() = (a_{pc}^{\dagger} P_{out}^{z} M_{out})^{\theta\theta(Npc^- 7)}$$

$$6$$
$$(a_{pc}^{\dagger} P_z^{r} M_z)^{\theta\theta(Npc^- 6)}$$

$$5$$
$$(\frac{a_{pc}^{\dagger} (a_r^{\dagger})^{N_y} \sqrt{N_r!}}{\sqrt{(N_r + N_y)!}})^{\theta\theta(Npc^- 5)}$$

$$4$$
$$(a_{pc}^{\dagger})^{1-\theta\theta(N_r)} ((a_{pc}^{\dagger})^{1-\theta(N_w)}(g())^{\theta(N_w)}a_w P_{pc}^{y} M_{pc})^{\theta\theta(N_r) \theta\theta(Npc^- 4)}$$

$$3$$
$$(a_{pc}^{\dagger} P_r^{x} M_r)^{\theta\theta(Npc^- 3)}$$

$$2$$
$$(a_{pc}^{\dagger} P_y^{in} M_y)^{\theta\theta(Npc^- 2)}$$

$$1$$
$$(a_{pc}^{\dagger} P_x^{in} M_x)^{\theta\theta(Npc^- 1)}$$

The program is

$$g()a_{pc}{}^{\dagger}M_{pc}\,(a_w{}^{\dagger})^{10}M_w\,|\,...>$$

where w is some memory location. We conjecture that any assembly language program using the previously specified instructions can be mapped to an algebraic representation – possibly with the use of additional memory for variables such as the counter variable seen above.

Using the algebraic constructs supporting non-sequential program execution we can create algebraic representations of assembly language programs. These programs have a definite input state and through the execution of the program they evolve into a definite output state --not a superposition of output states. Therefore they faithfully represent assembly language programs. On the other hand they are quantum in the sense that they use states and harmonic oscillator-like raising and lowering operators. The types of programs we are creating in this approach are "sharp" on the space of states. One input state evolves through the program's execution to one and only one output state with probability one.

These types of programs are analogous to free field theory in which incoming particles evolve without interaction to an output state containing the same particles.

In the next section we extend the ideas in this section to quantum programming where a variety of output states are possible – each with a certain probability of being produced.

10.5 Quantum Assembly Language™ Programs

In this section we will first look at a simplified quantum program that illustrates quantum effects but in actuality is a sum of deterministic assembly language programs mapped to algebraic equivalents. Consider a "quantum" program that is the sum of three ordinary programs $g_1()$, $g_2()$ and $g_3()$ of the type seen in the last section. Further let us assume the set of orthonormal states

$$|\,n,\,m,\,p,\,...>$$

that we saw in the previous sections with

$$<X\,|\,Y>\,=\delta_{XY}$$

where δ_{XY} represents a product of Kronecker δ functions in the individual values in memory of the $|\,X>$ and $|\,Y>$ states. Further let us assume

$$|\,n_1,\,m_1,\,p_1,\,...>\,=g_1()|\,...>$$
$$|\,n_2,\,m_2,\,p_2,\,...>\,=g_2()|\,...>$$
$$|\,n_3,\,m_3,\,p_3,\,...>\,=g_3()|\,...>$$

for some initial state of the quantum computer. Then

$$\alpha g_1() + \beta g_2() + \gamma g_3()| \ldots >$$

is a "quantum" program where α, β, and γ are constants such that

$$|\alpha|^2 + |\beta|^2 + |\gamma|^2 = 1$$

The quantum program produces the state $| n_1, m_1, p_1, \ldots >$ with probability $|\alpha|^2$, the state $| n_2, m_2, p_2, \ldots >$ with probability $|\beta|^2$, and the state $| n_3, m_3, p_3, \ldots >$ with probability $|\gamma|^2$.

We now have a quantum probabilistic computer. The programs $g_1()$, $g_2()$ and $g_3()$ are being executed in *parallel* in a quantum probabilistic manner.

Currently, the most feasible way of creating a Quantum Computer with current technology or reasonable extrapolations of current technology is to create a material which approximates a lattice with spins at each lattice site that we can orient electromagnetically at the beginning of a program. The execution of a program takes place by applying electromagnetic fields that have a time dependence specific to the computation. The electromagnetic fields implement a custom-tailored set of interactions between the spins in the material that simulates the calculation to be performed.

The interactions are specified with some Hamiltonian or some effective Hamiltonian and the initial state of the lattice spins evolves dynamically to some configuration that is then measured.

The Hamiltonians are normally specified using the space-time formalism that is a familiar part of Quantum Mechanics. A Hamiltonian specifies the time evolution of a system starting from an initial state. We can introduce an explicit time dependence in states by using the notation:

$$| \Psi(t) >$$

to denote the state of a Quantum Computer at time t. The general state of the computer at time t can be written as a superposition of the number representation states:

$$| \Psi(t) > = \sum_n f_n(t)| n_1, n_2, n_3, \ldots >$$

where n represents a set of values $n_1, n_2, n_3, \ldots$

The time evolution of the states can be specified using the Hamiltonian operator H as

$$| \Psi(t) > = e^{-iHt}| \Psi(0) >$$

With this Hamiltonian formulation we can imagine wishing to simulate a physical (or mathematical) process, defining a Hamiltonian that corresponds to the process, and then creating an experimental setup using a set of lattice spins in some material that implements the simulation. The experimental setup will prepare the initial state of the spins, create a fine tuned interaction that simulates the physics of the process, and then, after the system has evolved, will measure the state of the system at time t. Repeated performance of this procedure will determine the probability distribution associated with the final state of the Quantum Computer. The probability distribution is specified by $|f_n(t)|^2$ as a function of the sets of numbers denoted by n.

A simple example of a Hamiltonian that causes a Quantum Computer to evolve in a non-trivial way is:

$$H = \sum_{m=0}^{\infty} a_{m+1}{}^{\dagger} a_m$$

(This example was chosen partly because it has a form similar to a Virasoro algebra generator in SuperString Theory.) Let us assume the initial state of the Quantum Computer at $t = 0$ is

$$| 1, 0, 0, 0, \ldots >$$

that is, an initial value of 1 in the first word in memory and zeroes in all other memory locations. At time t the state of memory is:

$$n^{th} \text{ memory location}$$
$$\downarrow$$

$$| \Psi(t) > = \sum_{n=0}^{\infty} f_n(t)| 0, 0, \ldots, 1, 0, \ldots >$$

with

$$f_n(t) = (-it)^n/n!$$

using the power series expansion of the exponentiated Hamiltonian expression. The probability of finding the state

$$n^{th} \text{ memory location}$$
$$\downarrow$$

$$| 0, 0, \ldots, 1, 0, \ldots >$$

is

$$(t^n/n!)^2$$

At first glance the Hamiltonian approach is very different from the Quantum Assembly Language approach discussed above. However these approaches can be interrelated in special cases and (we conjecture) in the general case through sufficiently clever transformations. For example, the preceding Hamiltonian can be re-expressed as assembly language instructions

$$H = \sum_{m=0}^{\infty} (STORE\ (m+1))(ADD\ "1")(LOAD\ (m+1)) \cdot$$

$$\cdot (STORE\ m)\ (SUBTRACT\ "1")(LOAD\ m)$$

where a value is loaded into the register from memory location m and then 1 is added to the value in the register. The "1" expression represents a literal value one not a memory location. The parentheses around m+1 indicates it is the $(m+1)^{th}$ memory location – not the addition of one to the value at the mth location.

The preceding assembly language expression for H can be replaced with the algebraic representation expression:

$$H = \sum_{m=0}^{\infty} P_{m+1}{}^r\ M_{m+1}\ a_r{}^{\dagger}\ \frac{1}{\sqrt{(N_r + 1)}}\ P_r{}^{m+1} M_r P_m{}^r M_m a_r \sqrt{N_r} P_r{}^m M_r$$

This complex expression is not an improvement in one sense. The original Hamiltonian expression was much simpler. Its importance is the mapping that it embodies from a quantum mechanical Hamiltonian to an assembly language expression to an algebraic representation of the assembly language.

If we regard the value in the register as a "scratchpad" value as programmers often do, then we can establish a representation of $a_m{}^{\dagger}$ and a_m in terms of the algebraic representation of assembly language instructions.

$$a_m{}^{\dagger} \equiv P_m{}^r\ M_m\ a_r{}^{\dagger}\ \frac{1}{\sqrt{(N_r + 1)}}\ P_r{}^m\ M_r$$

and

$$a_m \equiv P_m{}^r\ M_m\ a_r\ \sqrt{N_r}\ P_r{}^m\ M_r$$

The power series expansion of the exponentiated Hamiltonian in the previous example is an example of the use of Perturbation Theory. The direct solution of a problem is often not feasible because of the complexity of the dynamics. Physicists have a very well developed theory for the approximate solution of these difficult problems called Perturbation Theory. Perturbation Theory takes an exact solution of a simplified version of the problem and then calculates corrections to that solution that approximate the exact solution of the problem. In the preceding example the initial state of the Quantum Computer represents a time-independent description of the Quantum Computer. The time-dependent description of the Quantum Computer which is the sought-for solution requires the evaluation of the result of the application of the exponentiated Hamiltonian to the initial state. For a small elapsed time, the exponential can be expanded in a power series and the application of the first few terms of the power series to the initial state approximates the actual state of the Quantum Computer. Thus we have a Perturbation Theory for the time evolution of the Quantum Computer expressed as an expansion in powers of the elapsed time.

10.6 Bit-Level Quantum Computer Language

In the previous section we examined a Quantum Assembly Language with words consisting of an infinite sets of bits. In this section we will examine the opposite extreme – a Quantum Computer Language with one-bit words. One can also create Quantum Computer Languages for intermediate cases such as 32-bit words.

A Bit-Level Quantum Computer Language can be represented with anti-commuting Fermi operators b_i and $b_i^\dagger$ for $i = 0, 1, 2, \ldots$ representing each bit location in the Quantum Computer's memory with the anti-commutation rules:

$$\{ b_i, b_j^\dagger \} = \delta_{ij}$$

$$\{ b_i, b_j \} = 0$$

$$\{ b_i^\dagger, b_j^\dagger \} = 0$$

where δ_{ij} is 1 if $i = j$ and zero otherwise. We will assume an (unrealistic) one-bit register with a pair of raising and lowering operators r and $r^\dagger$ for the register with the anti-commutation relations:

$$\{ r_i, r_j^\dagger \} = \delta_{ij}$$
$$\{ r_i, r_j \} = 0$$
$$\{ r_i^\dagger, r_j^\dagger \} = 0$$

The ground state of the computer is the state with the values at all bit memory locations set to zero. It is represented by the vector

$$| 0, 0, 0, \ldots > \equiv | 0 > \equiv \Phi_V$$

A typical state of the computer will be represented with a vector such as

$$| 1, 1, 1, \ldots > = r^\dagger b_0^\dagger\, b_1^\dagger \ldots | 0 >$$

with the first number being the value in the register, the second number the value at memory location 0, the third number the value at memory location 1, and so on.

A specified Quantum Computer state evolves as a Quantum Computer Program executes to a final computer state. A Bit-Level Quantum Computer Program can be represented as an algebraic expression in anti-commuting raising and lowering operators. The approach is similar to the approach seen earlier in this chapter for infinite-bit words using commuting operators.

10.6.1 Basic Operators of the Bit-Level Quantum Language

The key operators that are required for the algebraic representation of a Bit-Level Quantum Computer Language are:

Fetch the Value at a Memory Location (Number Operator)

$$N_m = b_m^\dagger b_m$$

For example,

$$N_m | \dots , 1, \dots > = | \dots , 1, \dots >$$

$\uparrow$

m^{th} memory location value

et the Value at Memory Location m to Zero

$$M_m = \quad (b_m)^{N_m}$$

The above expression for M_m is symbolic. The expression represents the following expression in which the operators are carefully ordered to avoid complications (c-numbers etc.) resulting from reordering.

$$M_m \equiv e^{N_m \ln b_m} = \sum_q \frac{(\ln b_m)^q N_m^q}{q!}$$

where the sum ranges from 0 to ∞. M_m becomes

$$M_m = 1 + (b_m - 1)N_m$$

using the identity $N_m = N_m^2$. When M_m is applied to a state it sets the value of the mth memory location to zero.

$$M_m | \dots , x, \dots > = \quad | \dots , 0, \dots >$$

$\uparrow$

m^{th} memory location value

Change Value at Memory Location m from 0 to Value at Location n

$$P_m{}^n = (b_m{}^\dagger)^{N_n}$$

The above expression for $P_m{}^n$ is also symbolic. The expression represents the following expression in which the operators are carefully ordered to avoid complications (c-numbers etc.) resulting from reordering.

$$P_m{}^n \equiv \sum_q \frac{(\ln b_m{}^\dagger)^q N_n{}^q}{q!}$$

where the sum over q ranges from 0 to ∞. Using the identity $N_m = N_m{}^2$ the expression for $P_m{}^n$ simplifies to:

$$P_m{}^n = 1 + (b_m - 1)N_m$$

When $P_m{}^n$ is applied to a state it changes the value of the mth memory location from zero to the value at the nth memory location.

$$P_m{}^n | \ldots, 0, \ldots, x, \ldots > = (b_m{}^\dagger)^x | \ldots, 0, \ldots, x, \ldots>$$

$$= | \ldots, x, \ldots, x, \ldots >$$

We can use the operators M_m and $P_m{}^n$ to express bit-wise assembly language instructions:

LOAD m – load the value at memory location m into the register

$$P_r{}^m M_r = (1 - N_r + b_r)(1 - N_m) + (N_r + b_r{}^\dagger)N_m$$

The first term on the right handles the case $N_m = 0$ and the second term on the right handles the case $N_m = 1$.

STORE m – store the value in the register at memory location m

$$P_m{}^r M_m = (1 - N_m + b_m)(1 - N_r) + (N_m + b_m{}^\dagger)N_r$$

The first term on the right handles the case $N_r = 0$ and the second term on the right handles the case $N_r = 1$.

ADD m   –   add the value at memory location m to the value in the register

$$(b_r^\dagger)^{N_m} = \sum_q \frac{(\ln b_r^\dagger)^q \, N_m^{\,q}}{q!}$$

$$= 1 + (b_r^\dagger - 1)N_m$$

If both the register and memory bit m have values of one then the application of this operator expression to the quantum state produces zero.

SUBTRACT m   – subtract the value at memory location m from the value in the register

$$(b_r)^{N_m} = \sum_q \frac{(\ln b_r)^q \, N_m^{\,q}}{q!}$$

$$= 1 + (b_r - 1)N_m$$

If the value in the register is zero and the value at location m is one the application of this operator produces zero.

MULTIPLY m   – multiply the value in the register by the value at memory location m

$$(b_r^\dagger)^{(N_m - 1)N_r} = \sum_q \frac{(\ln b_r^\dagger)^q \, (N_r)^q (N_m - 1)^q}{q!}$$

$$= 1 + (b_r - N_r)(1 - N_m)$$

Other assembly language instructions can be expressed in algebraic form as well.

The operator algebra that we have developed for a bit-wise Quantum Assembly Language™ or a Quantum Machine Language provides a framework for the investigation of the properties of Quantum Languages within an algebraic framework – a far simpler task than the standard quantum linguistic approaches.

10.7 Quantum High Level Computer Language Programs

The Quantum Assembly Language representation that we have developed earlier in this chapter forms a basis for high level Quantum Programming Languages. These languages are analogous to high level computer languages such as C or C++ or FORTRAN.

In ordinary computation a statement in a high level language such as

$$a = b + c;$$

in C programming is mapped to a set of assembly language by a C compiler. A simple mapping of the above C statement to assembly language would be

> LOAD ab
> ADD ac
> STORE aa

where aa is the memory address of a, ab is the memory address of b and ac is the memory address of c.

If we decide to define a High Level Quantum Computer Language then it would be natural to define it analogously in terms of a Quantum Assembly Language. A statement in the High Level Quantum Computer Language would map to a set of Quantum Assembly Language instructions.

For example, $a = b + c$ would map to the algebraic expression

$$P_{aa}{}^r \, M_{aa} \quad (a_r{}^\dagger)^{N_{ac}} \, \frac{\sqrt{N_r!}}{\sqrt{(N_r + N_{ac})!}} \, P_r{}^{ab} \, M_r$$

using the formalism developed earlier in this chapter to LOAD, ADD and STORE.

The definition of high level Quantum Computer Languages in this approach is straightforward. One can then imagine creating programs in these languages for execution on Quantum Computers just as ordinary programs are created for ordinary computers.

Another approach to higher level Quantum Computer Languages is to simply express them directly using raising and lowering operators – not in terms of Quantum Assembly Language instructions. For example the preceding $a = b + c;$ statement can be directly expressed as

$$(a_{aa}{}^\dagger)^{N_{ac}+N_{ab}} \, (a_{aa})^{N_{aa}} \, \frac{1}{\sqrt{(N_{ac} + N_{ab})!} \, \sqrt{N_{aa}!}}$$

Simple High Level Quantum Computer programs can be expressed as products of algebraic expressions embodying the statements of the program. These programs are sharp on the set of memory states taking an initial memory state that is an eigenstate of the set of number operators N_m into an output eigenstate of the number operators.

A general High Level Quantum Computer Program is a sum of simple High Level Programs. For example,

$$\alpha h_1() + \beta h_2() + \gamma h_3()| \dots >$$

where α, β, and γ are constants such that

$$|\alpha|^2 + |\beta|^2 + |\gamma|^2 = 1$$

The sum of simple programs $\alpha h_1() + \beta h_2() + \gamma h_3()$ produces the state $| n_1, m_1, p_1, \dots >$ with probability $|\alpha|^2$, the state $| n_2, m_2, p_2, \dots >$ with probability $|\beta|^2$, and the state $| n_3, m_3, p_3, \dots >$ with probability $|\gamma|^2$.

An initial eigenstate of the number operators is transformed into an output state that is a superposition of number operator eigenstates. In this case we use probabilities to specify the likelihood that a given output eigenstate will be found when the output state is measured.

A Hamiltonian can also be used to specify the time evolution of a system starting from an initial state. Using the notation:

$$| \Psi(t) >$$

to denote the state of a Quantum Computer at time t the general state of a computer at time t can be written as a superposition of number representation states:

$$| \Psi(t) > = \sum_n f_n(t)| n_1, n_2, n_3, \dots >$$

where n represents a set of values $n_1, n_2, n_3, \dots$

The time evolution of the states can be specified using the Hamiltonian operator H as

$$| \Psi(t) > = e^{-iHt}| \Psi(0) >$$

A simple example of a Hamiltonian that causes a Quantum Computer to evolve in a non-trivial way is:

$$H = \sum_{m=0}^{\infty} (a_{m+2}{}^\dagger)^{N_{m+1}+N_m} (a_{m+2})^{N_{m+2}} \frac{1}{\sqrt{(N_{m+1}+N_m)!} \sqrt{N_{m+2}!}}$$

This Hamiltonian is based on the a = b + c statement above. This Hamiltonian generates a complex superposition of states as time evolves. More complex Hamiltonians equivalent to programs with several statements can be easily constructed.

10.8 Quantum C Language

One of the most important computer languages is the C programming language developed at Bell Laboratories in the 1970's. The original version of version of the C language was a remarkable combination of low level (assembly language-like) features and high level features like the mathematical parts of FORTRAN. The variables in the language were integers stored in words just as we saw in the earlier examples in this chapter. (There were several other types of integers as well – a complication that we will ignore.)

Using the ideas seen in the earlier sections of this chapter it is easy to develop algebraic equivalents for most of the constructs of the C language and thus create a Quantum C Language. An important element that must be added to the previous development is to introduce the equivalent of pointers. Simply put pointers are variables that have the addresses of memory locations as their values. The C language has two important operators for pointer manipulations:

Operator	Role	Example
&	Fetch an address	ptr = &x;
*	Fetch/set the value at an address	z = *ptr;
		*ptr = 99;

The & operator of C fetches the address of a variable in memory. The example shows a pointer variable ptr being set equal to the address of the x variable. The * (dereferencing) operator can fetch the value at a memory location. The first * example illustrates this aspect: the variable z is set equal to the value at the memory location specified by the pointer variable ptr. The * operator can also be used to set the value at a memory location as illustrated by the second * example. In this example the value 99 is placed at the memory location (address) specified by the ptr pointer variable.

These operators can be implemented in the algebraic representation of the Quantum C Language in the following way:

$$\& \quad \Leftrightarrow \quad [A, \;]$$

where $A = \Sigma \, m(a_m - a_m^\dagger)$ with the sum from 0 to ∞. If we apply the operator to a raising or lowering operator we obtain its address

$$\& \; a_m^\dagger \; = [A, a_m^\dagger] = m = \& \; a_m = [A, a_m]$$

The equivalent of the * operator is actually a pair of operator expressions. To fetch the value at a memory location we use

$$*m \equiv N_m$$

To set the value to X at a memory location m we use a more complex C language representation:

$$*m = X;$$

An equivalent algebraic expression is:

$$*m(X) \equiv (a_m{}^\dagger)^X \, (a_m)^{N_m} \; \frac{1}{\sqrt{X!} \; \sqrt{N_m!}}$$

$*m(X)$ is a functional notation. So a = b + c can be rewritten as a "pointer" algebraic expression as:

$$(a_{aa}{}^\dagger)^{*ac + *ab} \, (a_{aa})^{*aa} \; \frac{1}{\sqrt{(*ac + *ab)!} \; \sqrt{*aa!}}$$

Or more compactly using the functional notation as

$$*aa(*ab + *ac)$$

The Quantum C Language could be used to define Hamiltonians for a Quantum Computer. Other languages such as Java™, C++, lisp and so on also have Quantum analogues which may be defined in a similar way.

Appendix A. Hypercomplex Cosmology Connection to the Unified SuperStandard Theory (UST and NEWUST)

The author's UST and NEWUST theories originated in the past twenty years from the Standard Model of Particles with SU(2)⊗U(1)⊗SU(3) internal symmetries combined with Two-Tier Quantum and PseudoQuantum Field Theory (GiFT).

Noting the presence of conserved particle numbers, and the presence of at least three fermion generations, we introduced the U(4) Generation Groups and the U(4) Layer Groups together with four layers of four generation of "Normal" fermions necessitated by Generation Group and Layer group symmetries at the point of the Big Bang prior to symmetry breaking. The Dark sector had a corresponding set of four layers of four generations of "Dark" fermions. The result was the Unified SuperStandard Theory (UST) with the symmetry:

$$\{[SU(2)\otimes U(1)\otimes SU(3)]^2\otimes U(4)^4\}^4$$

supplemented with an additional Strong Interaction U(1) group. Space-time had four dimensions.

In the Fall, 2019 the author discovered that a hypercomplex-based theory (Octonion Cosmology space with Cayley-Dickson number n = 3) that he constructed and named QUeST had the same internal symmetries as UST with the addition of $U(1)^8$. The addition of $U(1)^8$ indicates that the Strong Interactions in the theory are broken Strong SU(4).

QUeST's internal symmetries, which could be based on a 16 × 16 dimension array of dimensions, was

$$[SU(2)\otimes U(1)\otimes SU(3)\otimes U(1)]^8\otimes U(4)^{16}$$

During 2020 the author developed an hypercomplex space spectrum for both universes and Megaverses, and other spaces. The spectrum was shown to arise from a generation mechanism whereby fermion-antifermion annihilation in a higher space produced an instance of a lower space. A critical part of the derivation of the hypercomplex spectrum was the realization that even space-time dimension spinor arrays are composed of Cayley number rows and columns. Spinor arrays of annihilating fermion-antifermion pairs were shown to generate the arrays of dimensions of subspace instances.

Analyzing the spinor arrays the author noted that the dimension array could be viewed as composed of 64 dimension subblocks, which were further subdivided into 16 dimension subblocks.

The subblock structuring, using the known contents of the Standard Model plus Generation and Layer groups for guidance, gave the dimension array structure containing 4 × 4 subblocks in Figs. A..1 and A.2.

Thus there was a *most* satisfactory match between UST and QUeST with the only significant difference being the space-time: four octonion (complex quaternion) coordinates for QUeST and four real space-time coordinates for UST. This difference was resolved in NEWUST and NEWQUeST. See the Connection groups in Fig. A.3.

Figs. A.4 and A.5 show the detailed group and fermion structures.

The form of the square spinor arrays generated by fermion-antifermion annihilation gives 64 dimension blocks and 16 dimension blocks as well as 32 dimension composite blocks that are evidenced in the NEWQUeST fermion spectrums and internal symmetry group structure.

A.1 Dimensions of Symmetries and Coordinates

Since we see only real dimensions in Reality, we transferred 28 QUeST dimensions from space-time to $U(2)^7$ internal symmetry dimensions. The set of internal symmetries was increased by $U(2)^7$, which we call Connection Groups. Each Connection group specifies interactions between corresponding fermions (e with e, q with q, and so on) in separate layers and between Normal and Dark fermions. The connections between the various blocks of fermions are shown in Fig. A.3. *We implement the very practical rule that all blocks must be connected by interactions or they would not be of physical interest. A totally isolated block effectively does not exist physically (except possibly for gravitation effects).*

The interactions of the Connection groups must be very weak and/or their gauge bosons must be very massive.

The addition of the Connection Groups and the reduction of space-time dimensions accordingly results in NEWQUeST and NEWUST as summarized below.

Note: the Generation, Layer, and Connection groups are all badly broken. Their vector bosons must be very massive since they have not been detected in experiments.

A.1.1 Internal Symmetries

The groups are ElectroWeak SU(2)⊗U(1), Strong SU(3), Generation Group U(4), Layer Group U(4), and U(2) and U(4) Connection groups obtained by transfer from space-time coordinates (See Blaha 2012c). The SU(3)⊗U(1) symmetries may be a broken SU(4) symmetries. The internal symmetries for the theories are:

UST

$$[SU(2)\otimes U(1)\otimes SU(3)]^8 \otimes U(4)^{16} \tag{A.1}$$

QUeST

$$[SU(2)\otimes U(1)\otimes SU(3)\otimes U(1)]^8 \otimes U(4)^{16} \tag{A.2}$$

NEWQUeST

$$[SU(2)\otimes U(1)\otimes SU(3)\otimes U(1)]^8 \otimes U(4)^{16} \otimes U(2)^7 \tag{A.3}$$

The only change is in internal symmetries: Twenty-eight real dimensions transferred from space-time coordinates to Connection group $U(2)^7$ internal symmetry.

NEWUST

$$[SU(2)\otimes U(1)\otimes SU(3)\otimes U(1)]^8 \otimes U(4)^{16} \otimes U(2)^7 \qquad (A.4)$$

The only change in internal symmetries: Twenty-eight real dimensions added for $U(2)^7$ Connection group internal symmetry.

A.1.2 Space-Time Coordinates

UST

>Four real space-time coordinates.

QUeST

>Four octonion (complex quaternion) coordinates.

NEWUST

>Four real space-time coordinates. No change from UST space-time.

NEWQUeST

>Four real space-time coordinates. The six coordinates in the $n = 4$ octonion space were lowered to four space-time coordinates with two coordinates transferred to Connection groups.

The only change is in space-time coordinates: Fourteen dimensions transferred from QUeST space-time coordinates to Connection group $U(2)^7$ internal symmetry.

A.2 Fundamental Fermion Spectrum

There are 256 fundamental fermions in NEWQUeST and NEWUST. Conceptually their structure can be viewed as an extrapolation of the known three generations of The Standard Model. For good reason (U(4) Generation groups) a fourth generation was indicated and a corresponding Dark sector of similar structure was added. In addition, because of the need for Layer groups, the overall structure consisted of four copies of this layer (due to U(4) Layer groups).

Correspondingly, each layer also has its own set of internal symmetry gauge groups to limit mixing between the layers to Layer group interactions and Connection group interactions.

Fig. A.1 shows the structure of the NEWQUeST/NEWUST fermions. The blocks are 4×4 reflecting the origin of the NEWQUeST/NEWUST space (universe) instance from Megaverse fermion-antifermion annihilation. The spinor analysis of their spinor arrays yields a 16 dimension block structure. The 64 dimension fermion layers reflect the 64 dimension structuring of the Megaverse obtained from its creation by fermion-antifermion creation in the Maxiverse.

A.3 Total Dimensions

The total of internal symmetry and space-time dimensions is 256 in all four theories listed above. It is based on the 16×16 dimension array of the Cayley-Dickson number n = 3 space of the Hypercomplex spectrum.

A.4 Pattern of Internal Symmetries

The NEWQUeST dimension array for internal symmetries is subdivided into four layers of 56 dimensions—just as in NEWUST (and UST). Fig. A.2 displays the layers using SU(4) in place of SU(3)⊗U(1). Each layer has a block of 28 dimensions for Normal and 28 dimensions for Dark sectors. There are also seven U(2) Connection groups plus four real-valued space-time coordinates bringing the NEWQUeST total to 256 dimensions.= 4*56 + 28 + 4 = 256 dimensions. The Connection groups are shown in Fig. A.3.

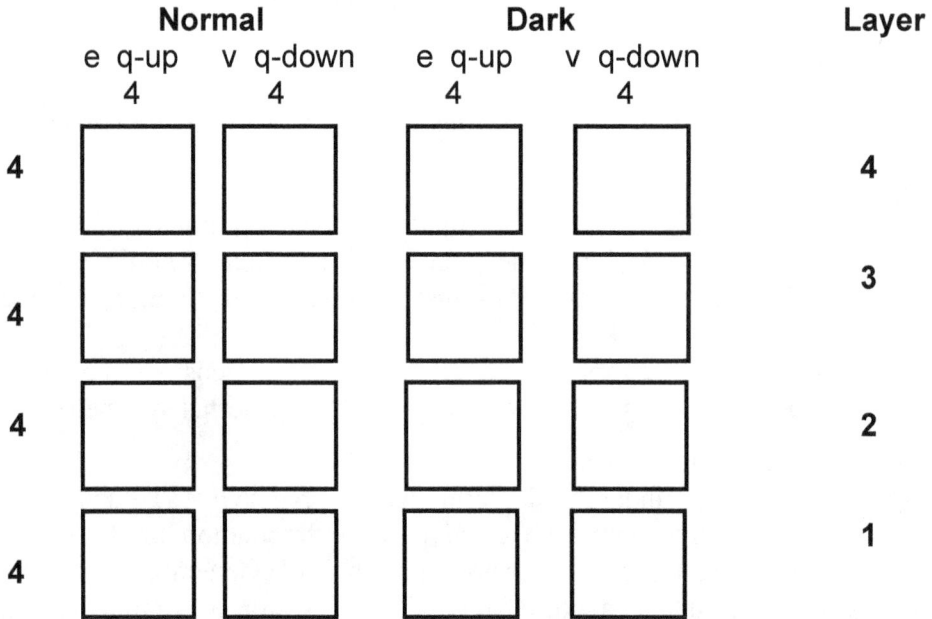

Figure A.1. Block form of a 16×16 NEWQUeST/NEWUST fermion array with each block row corresponding to one layer. Each block contains four generations of fermions. The result is 4×4 blocks. The label e q-up indicates a charged lepton – up-type quark pair, v q-down indicates a neutral lepton – down-type quark pair, and so on. The blocks can be viewed as SU(3)⊗U(1) or broken SU(4) blocks.

Layers	NORMAL		DARK	
	4	4	4	4
4	SU(2)⊗U(1)⊗SU(3)⊗U(1) 4 Space-time Dimensions	Generation + Layer Groups	SU(2)⊗U(1)⊗SU(3)⊗U(1) 4 Space-time Dimensions	Generation + Layer Groups
4	SU(2)⊗U(1)⊗SU(3)⊗U(1) 4 Space-time Dimensions	Generation + Layer Groups	SU(2)⊗U(1)⊗SU(3)⊗U(1) 4 Space-time Dimensions	Generation + Layer Groups
4	SU(2)⊗U(1)⊗SU(3)⊗U(1) 4 Space-time Dimensions	Generation + Layer Groups	SU(2)⊗U(1)⊗SU(3)⊗U(1) 4 Space-time Dimensions	Generation + Layer Groups
4	SU(2)⊗U(1)⊗SU(3)⊗U(1) 4 Space-time Dimensions	Generation + Layer Groups	SU(2)⊗U(1)⊗SU(3)⊗U(1) 4 Space-time Dimensions	Generation + Layer Groups

Figure A.2.. Four layers of Internal Symmetry groups in NEWQUeST and NEWUST (omitting Connection Groups) showing the 4 by 4 subblocks. The groups in each layer are independent of those in other layers. The groups in each subblock of each layer are independent of those in the other subblocks. Each subblock contains 16 dimensions. The block dimensions furnish fundamental representations for the groups listed. The entire set of blocks contains 256 dimensions. Each layer contains 56 internal symmetry dimensions. The first two columns are for the "Normal" sector. The last two columns are for the "Dark" sector (although most of the Normal sector is Dark observationally at present.) This figure also holds for UST with the addition of U(1) groups. The eight sets of 4 real dimension space-times combine to give a 4 real dimension space-time and seven U(2) Connection groups.

Connection Group Applied to Fermions

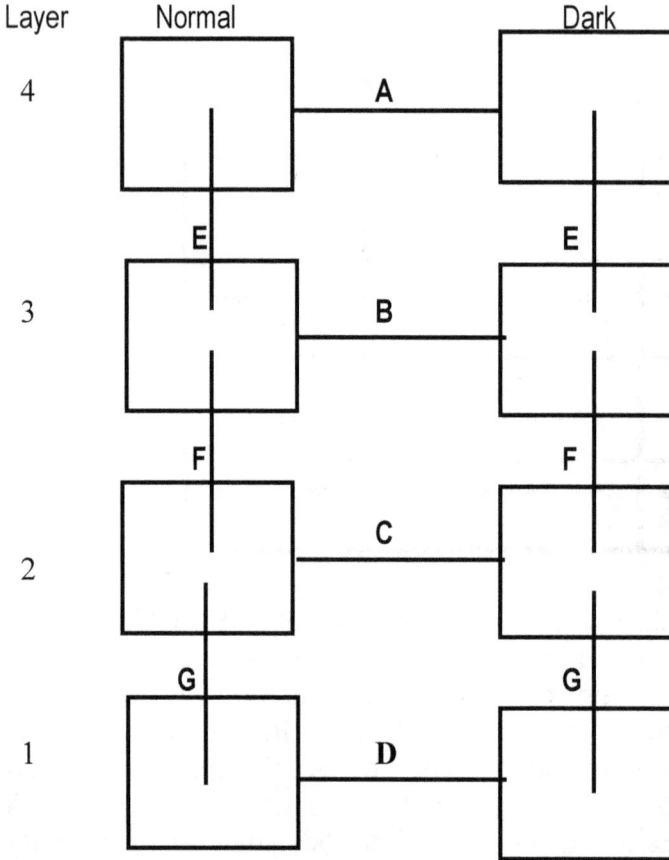

Figure A.3. The seven U(2) Connection groups (shown as 10 lines) between the eight NEWQUeST/NEWUST blocks. Connection groups are obtained by transfering 28 dimensions from QUeST space-time to internal symmetries with the consequent reduction of the space-time from four octonion (complex quaternion) coordinates to four real coordinates. The Connection groups generate rotations and interactions between corresponding fermions and vector bosons of each pair of blocks. The Normal and Dark sector U(2) vertical connections above (E, F, G) represent the same U(2) groups.

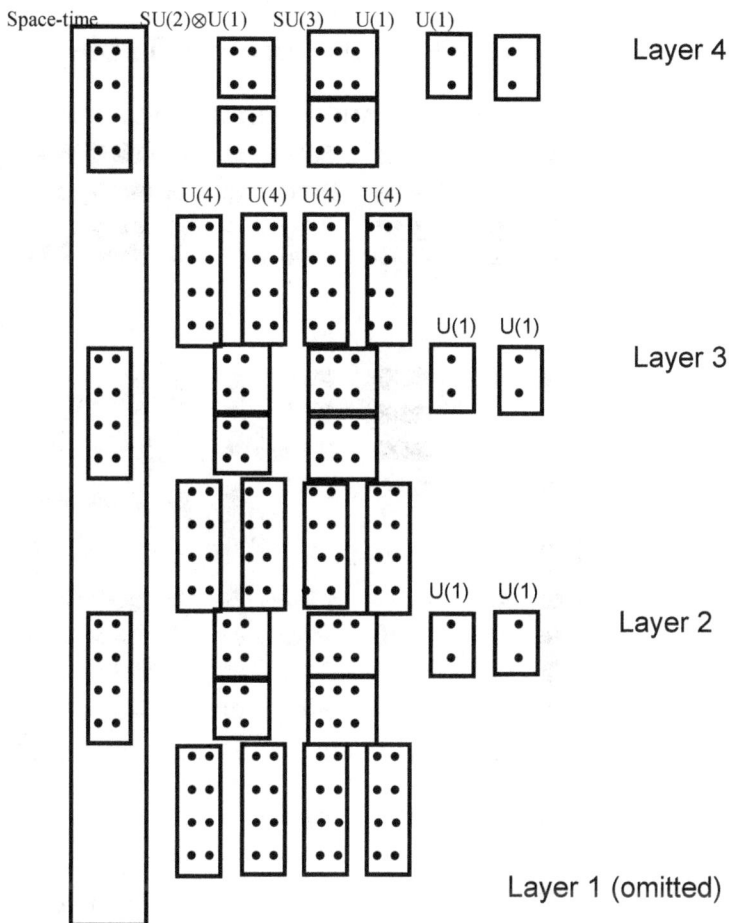

Figure A.4. Three of the four layers of QUeST internal symmetry groups (and space-time) for Cayley-Dickson space 3. Layer 1 which has an identical form was omitted due to "page space" limitations. Note the left column of blocks combine to specify a 4 dimension octonion space-time. Note each layer has 64 dimensions.

The Fermion Periodic Table

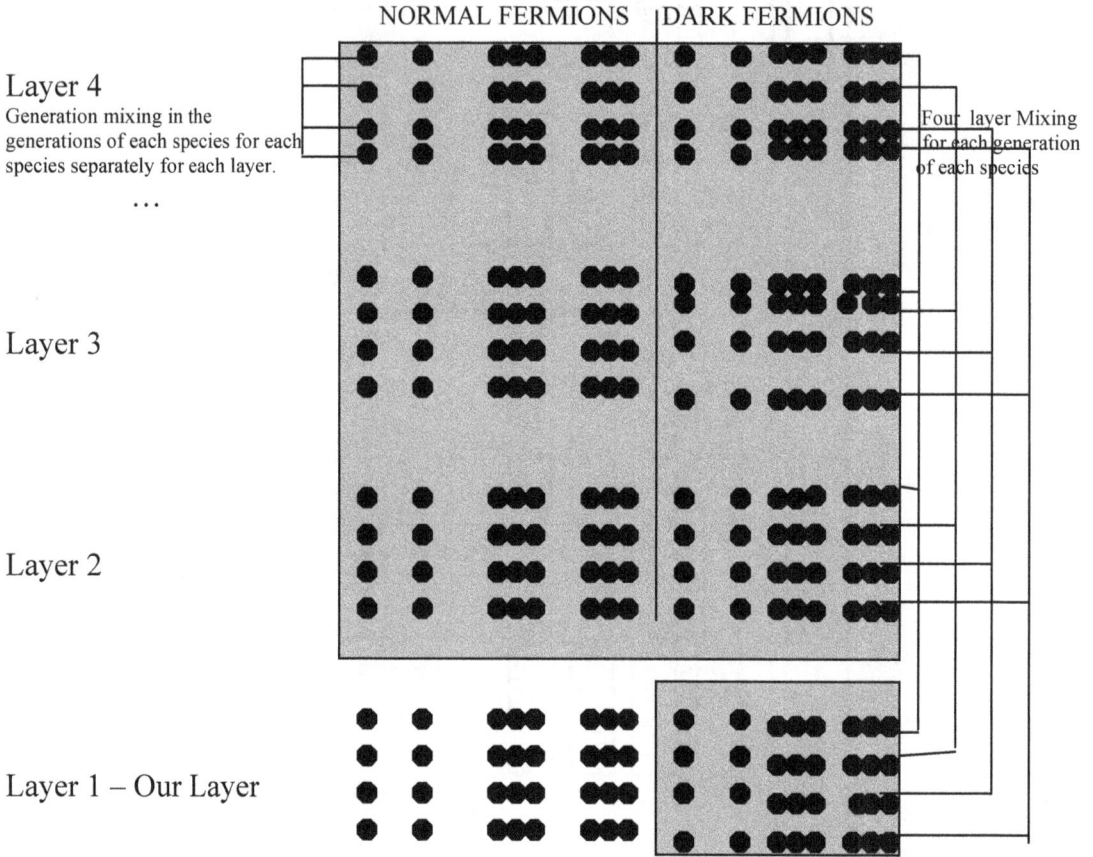

Figure A.5. Fermion particle spectrum and partial examples of the pattern of mass mixing of the Generation group and of the Layer group. Unshaded parts are the known fermions with an additional, as yet not found, 4th generation. The lines on the left side (only shown for one layer) display the Generation mixing within each layer. The Generation mixing occurs within each layer using a separate Generation group for each layer. The lines on the right side show Layer group mixing (for Dark matter) with the mixing among all four layers for each of the four generations individually. There are four Layer groups for Normal matter and four Layer groups for Dark matter.. There are 256 fundamental fermions. QUeST and UST have the same fermion spectrum.

Appendix B. NEWUTMOST for the Megaverse

The NEWUTMOST theory presented in earlier books corresponds to the space of Cayley-Dickson number 4 (Blaha space number 6). It describes a Megaverse (Multiverse). This chapter outlines its features.

The NEWUTMOST dimension array has the size 32×32. It has 1024 dimensions.

B.1 NEWUTMOST Space and Internal Symmetries

An octonion contains eight dimensions. A complex octonion contains sixteen dimensions. A *quaternion octonion* contains 32 dimensions. Fig. B.1 depicts the 32 dimension complex octonion space as a 32×32 array of dimensions. It uses a "dot" or pebble • to represent a dimension. The dimensions of the space are not assigned physically until they are mapped to internal symmetry group fundamental representation dimensions and space-time dimensions. Rather than create a cumbersome coordinate-based notation we choose to use •'s.

Figure B.1. The 32 quaternion octonion dimension NEWUTMOST array. It is a 32×32 dimension array of •'s. It has 1024 dimensions. It is grouped into four layers in rows of eight.

The repetitive pattern of groups seen in QUeST leads us to assume that NEWUTMOST has a similar repetitive pattern. We will use a four layer format for the 32×32 array of dimensions. Each layer consists of 8 rows of Fig. B.1. Each layer can be put in a form analogous to Fig. C.4 (and to Fig. C.9). See Fig. B.3.

We map between dimensions and fundamental group representations. We use the maps in Table B.2 to set up the group ↔ dimension map, bearing in mind the group representations of the Standard Model:

U(4)	↔ 8 real dimensions
U(2)	↔ 4 real dimensions
SU(3)	↔ 6 real dimensions
U(1)⊗SU(2)	↔ 4 real dimensions
U(1)	↔ 2 real dimensions

Table B.2. Map between fundamental representations and their dimensions.

Figs. B.3 and B.4 show the content of one NEWUTMOST layer. The four layers of NEWUTMOST are four copies of Fig. B.3.[13] The separation of the set of dimensions is accomplished by following the procedure given earlier.[14]

Fig. B.5 shows the 4×4 blocks in the four layers (each in two rows) of the 32×32 dimension NEWUTMOST array. The 4×4 blocks are within the four block 8×8 sections for each pair: Normal+Dark1, Dark2+Dark3, Dark4+Dark5 and Dark6+Dark7. In total they form the $32 \times 32 = 1024$ NEWUTMOST dimension array.

---

[13] The Layer groups are U(4) groups. They mix the generations of each of the top four layers, generation by generation, separately from the Layer groups mixing the lower four layers. This feature enables NEWQUeST universes to be generated from either the top four layers or the lower four layers.

[14] The separation of the dimensions into the subgroup factors' representation can be implemented as group transformations and definitions using standard group theoretic methods. A more formal method for extracting the subgroup content of representations uses a symmetric group analysis of U(n) representation characters. See S. Blaha, J. Math. Phys. **10**, 2156 (1969) for a detailed discussion of this approach.

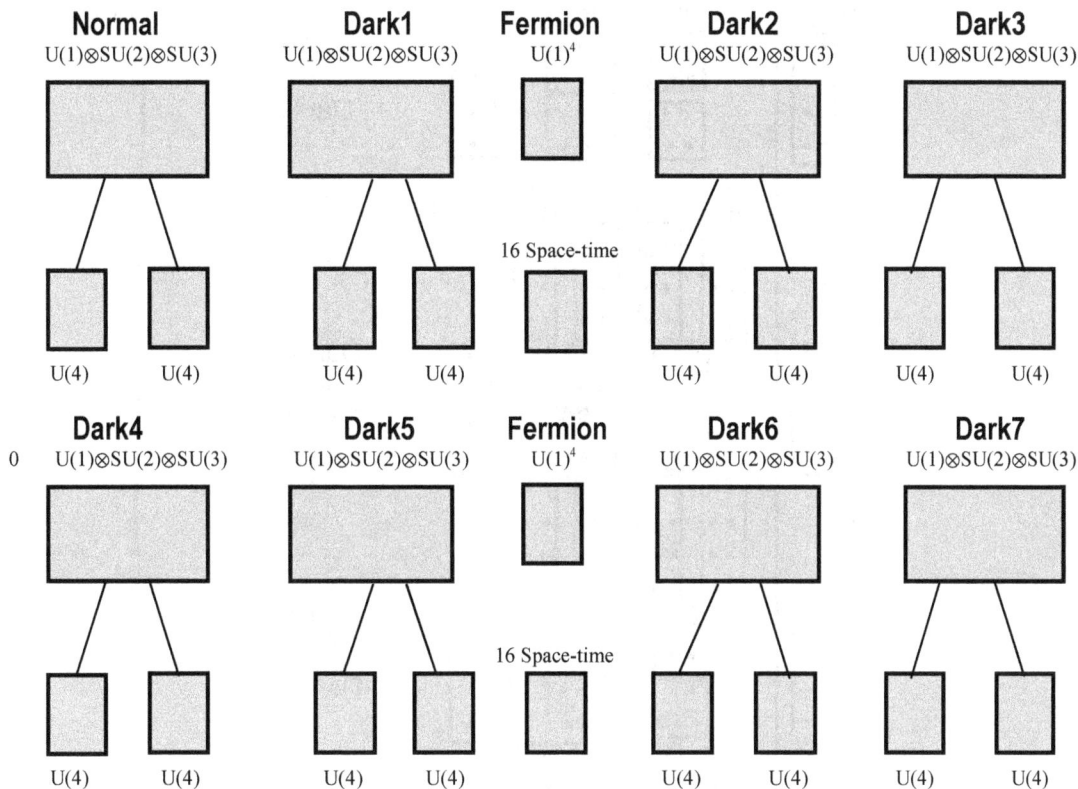

Figure B.3. The internal symmetry groups of *one layer* (consisting of 8 rows in Fig. B.1) of the four layers of 32 × 32 dimension NEWUTMOST. The other three layers are copies of the this layer. Note the Fermion $U(1)^4$ groups.

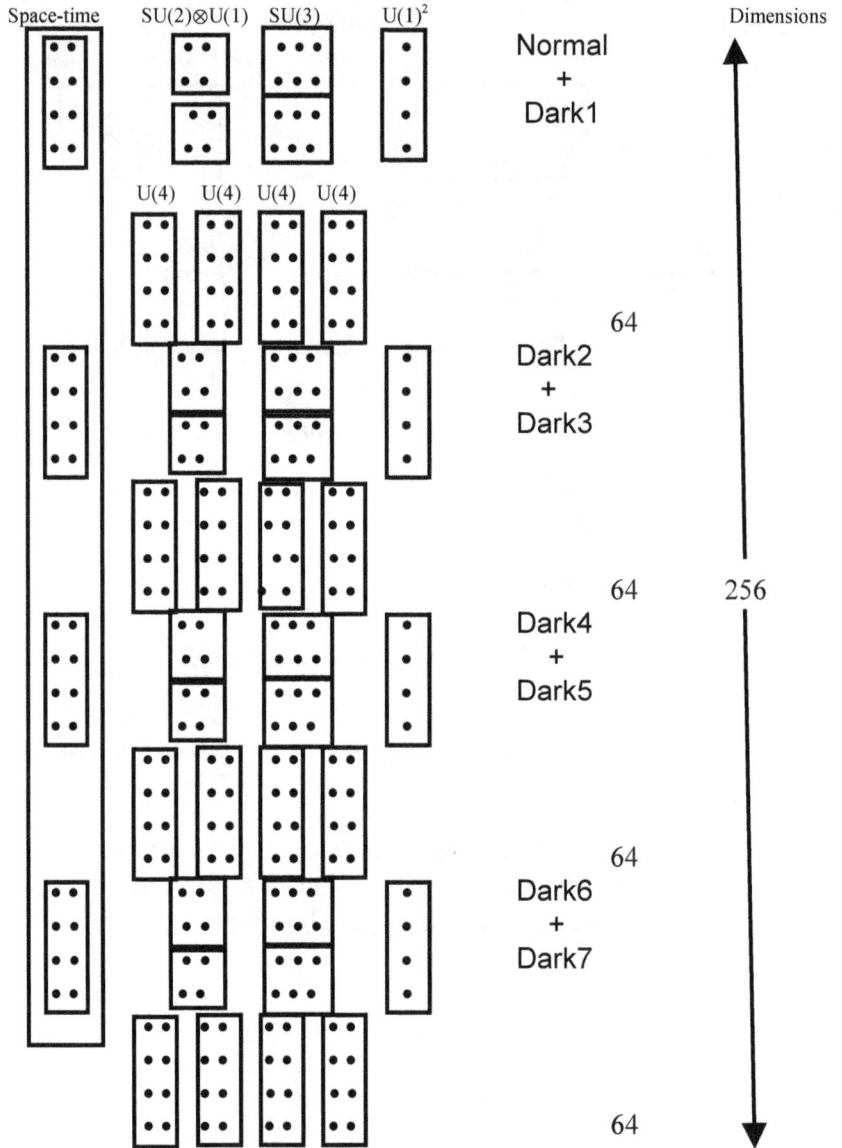

Figure B.4 The *first* of the four layers of NEWUTMOST dimensions with boxes around sets of dimensions for fundamental group representations.

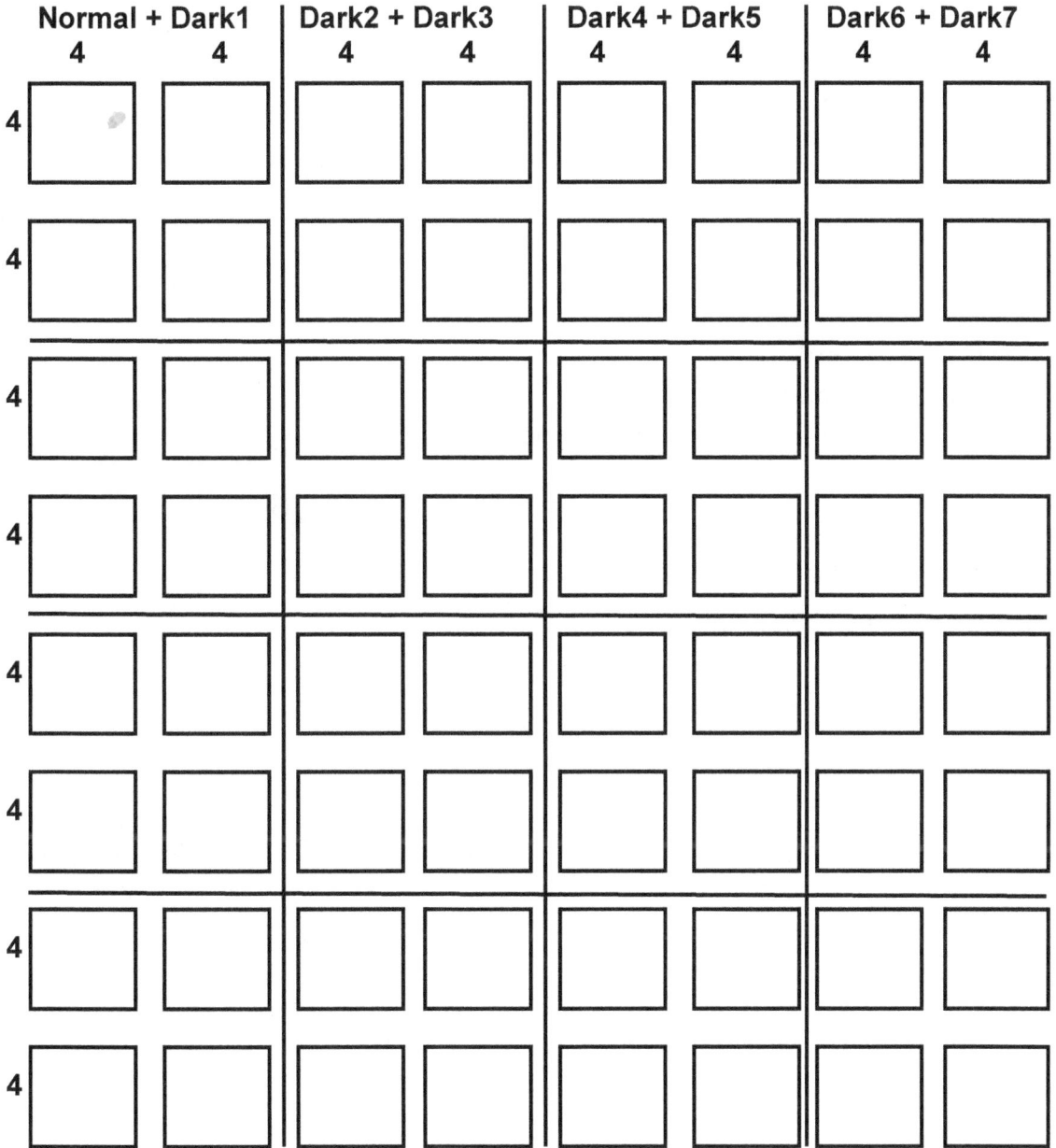

Figure B.5. Four layers (each in two rows) in the 32 × 32 dimension NEWUTMOST array composed of 4 × 4 blocks, which are within the four block 8 × 8 sections for each pair: Normal+Dark1, Dark2+Dark3, Dark4+Dark5 and Dark6+Dark7. In total they form the 32 × 32 = 1024 NEWUTMOST dimension array.

The NEWUTMOST dimension array is partitioned into $4 \times 4 = 16$ dimension blocks within $8 \times 8 = 64$ dimension blocks.[15] Fig. B.5 displays the three types of 16 dimension blocks:

A. $U(4) \otimes U(4)$
B. $U(1) \otimes SU(2) \otimes U(1) \otimes SU(2)$
C. $SU(3) \otimes SU(3) \otimes U(1) \otimes U(1) \otimes (\text{Space-time})$.

These blocks are repeated throughout the four layers. *NEWUTMOST has 32 blocks of type A, 16 blocks of type B, and 16 blocks of type C.* The QUeST dimension structure (Fig. 2.4) results.

B.2 NEWUTMOST Symmetries

B.2.1 NEWUTMOST Internal Symmetry Groups

The NEWUTMOST internal symmetry group initially is

$$[SU(2) \otimes U(1) \otimes SU(3)]^{32} \otimes U(4)^{64} \otimes U(1)^{32} \tag{B.1}$$

It changes after transformations to Connection Groups.

B.2.2 NEWUTMOST Space-Time

The space-time of NEWUTMOST is initially 8 complex octonion dimensions or 128 dimensions. These dimensions are transformed to one 6 dimension space-time plus 122 dimensions for Connection Group symmetries.

B.2.3 NEWUTMOST Fermion Spectrum

Given the form of the internal symmetries in NEWUTMOST we can determine the fermions in the fundamental group representations as shown in Fig. B.6. The set of 1024 dimensions determines the set of fermions in fundamental representations. The fermions are ordered by layer.

B.3 NEWUTMOST Fermions

Given the form of the internal symmetries in NEWUTMOST we can determine the fermions in the fundamental group representations as shown in Fig. B.6.

---

[15] The $8 \times 8 = 64$ dimension blocks arise as the footprint of the urfermion-antiurfermion annihilation in the Maxiverse that generates an UTMOST Megaverse instance. See section 8.3.

NEWUTMOST Fermion Array

Normal	Dark1	Dark2	Dark3	Dark4	Dark5	Dark6	Dark7

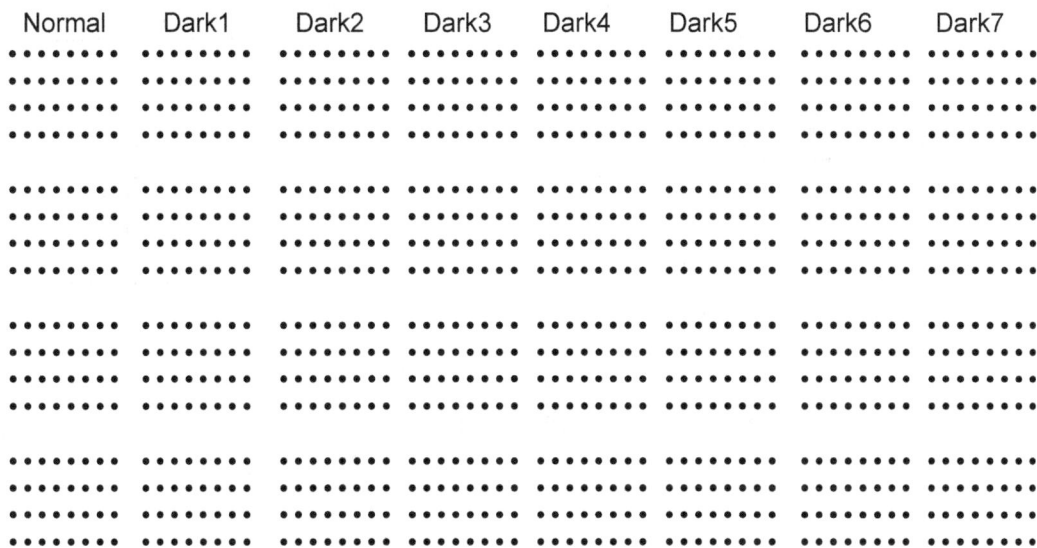

Figure B.B. Spectrum of NEWUTMOST fermions in a 16×64 format. Each fermion is represented by a •..Each set of eight •.'s represents a charged lepton, a neutral lepton, three up-type quarks, and three down-type quarks. There are eight sets of four species in four generations which are in turn in 4 layers. There are 1024 fundamental fermions taking account of quark triplets.

In Appendix A we outlined possible patterns of subspaces of NEWQUeST. One choice of pattern is based on 4×4 blocks of dimensions, assembled into 8×8 blocks of dimensions containing four 4×4 blocks, assembled in four layers.

Fig. B.7 shows the possible implications of this arrangement for NEWUTMOST fermions. The 4×4 fermion blocks contain either four generations of charged leptons and up-quarks, or four generations of neutral leptons and down-quarks.

The grouping of a lepton and three quarks in both cases creates a similarity to time and spatial coordinates respectively suggesting a broken Lorentz group-like structure or a possible SU(4) broken symmetry.

Normal				Dark1				Dark2				Dark3			
e q-up		v q-down		e q-up		v q-down		e q-up		v q-down		e q-up		v q-down	
4		4		4		4		4		4		4		4	

Figure B.7. Block form of the 32 × 32 NEWUTMOST fermion array with each row corresponding to *half of an NEWUTMOST layer*. Thus 8 × ½ = 4 layers results. Each block contains four generations of fermions. The result is sixty-four 4 × 4 blocks. The label e q-up indicates a charged lepton – up-type quark pair, v q-down indicates a neutral lepton – down-type quark pair, and so on. *The form displayed here explains why generations come in fours.*

REFERENCES

Akhiezer, N. I., Frink, A. H. (tr), 1962, *The Calculus of Variations* (Blaisdell Publishing, New York, 1962).

Bjorken, J. D., Drell, S. D., 1964, *Relativistic Quantum Mechanics* (McGraw-Hill, New York, 1965).

Bjorken, J. D., Drell, S. D., 1965, *Relativistic Quantum Fields* (McGraw-Hill, New York, 1965).

Blaha, S., 1995, *C++ for Professional Programming* (International Thomson Publishing, Boston, 1995).

_____, 1998, *Cosmos and Consciousness* (Pingree-Hill Publishing, Auburn, NH, 1998 and 2002).

_____, 2002, *A Finite Unified Quantum Field Theory of the Elementary Particle Standard Model and Quantum Gravity Based on New Quantum Dimensions™ & a New Paradigm in the Calculus of Variations* (Pingree-Hill Publishing, Auburn, NH, 2002).

_____, 2004, *Quantum Big Bang Cosmology: Complex Space-time General Relativity, Quantum Coordinates,™ Dodecahedral Universe, Inflation, and New Spin 0, ½, 1 & 2 Tachyons & Imagyons* (Pingree-Hill Publishing, Auburn, NH, 2004).

_____, 2005a, *Quantum Theory of the Third Kind: A New Type of Divergence-free Quantum Field Theory Supporting a Unified Standard Model of Elementary Particles and Quantum Gravity based on a New Method in the Calculus of Variations* (Pingree-Hill Publishing, Auburn, NH, 2005).

_____, 2005b, *The Metatheory of Physics Theories, and the Theory of Everything as a Quantum Computer Language* (Pingree-Hill Publishing, Auburn, NH, 2005).

_____, 2005c, *The Equivalence of Elementary Particle Theories and Computer Languages: Quantum Computers, Turing Machines, Standard Model, Superstring Theory, and a Proof that Gödel's Theorem Implies Nature Must Be Quantum* (Pingree-Hill Publishing, Auburn, NH, 2005).

_____, 2006a, *The Foundation of the Forces of Nature* (Pingree-Hill Publishing, Auburn, NH, 2006).

_____, 2006b, *A Derivation of ElectroWeak Theory based on an Extension of Special Relativity; Black Hole Tachyons; & Tachyons of Any Spin.* (Pingree-Hill Publishing, Auburn, NH, 2006).

_____, 2007a, *Physics Beyond the Light Barrier: The Source of Parity Violation, Tachyons, and A Derivation of Standard Model Features* (Pingree-Hill Publishing, Auburn, NH, 2007).

_____, 2007b, *The Origin of the Standard Model: The Genesis of Four Quark and Lepton Species, Parity Violation, the ElectroWeak Sector, Color SU(3), Three Visible Generations of Fermions, and One Generation of Dark Matter with Dark Energy* (Pingree-Hill Publishing, Auburn, NH, 2007).

_____, 2008a, *A Direct Derivation of the Form of the Standard Model From GL(16)* (Pingree-Hill Publishing, Auburn, NH, 2008).

_____, 2008b, *A Complete Derivation of the Form of the Standard Model With a New Method to Generate Particle Masses Second Edition* (Pingree-Hill Publishing, Auburn, NH, 2008)

_____, 2009, *The Algebra of Thought & Reality: The Mathematical Basis for Plato's Theory of Ideas, and Reality Extended to Include A Priori Observers and Space-Time Second Edition* (Pingree-Hill Publishing, Auburn, NH, 2009).

_____, 2010a, *Operator Metaphysics: A New Metaphysics Based on a New Operator Logic and a New Quantum Operator Logic that Lead to a Mathematical Basis for Plato's Theory of Ideas and Reality* (Pingree-Hill Publishing, Auburn, NH, 2010).

_____, 2010b, *The Standard Model's Form Derived from Operator Logic, Superluminal Transformations and GL(16)* (Pingree-Hill Publishing, Auburn, NH, 2010).

_____, 2010c, *SuperCivilizations: Civilizations as Superorganisms* (McMann-Fisher Publishing, Auburn, NH, 2010).

_____, 2011a, *21st Century Natural Philosophy Of Ultimate Physical Reality* (McMann-Fisher Publishing, Auburn, NH, 2011).

_____, 2011b, *All the Universe! Faster Than Light Tachyon Quark Starships & Particle Accelerators with the LHC as a Prototype Starship Drive Scientific Edition* (Pingree-Hill Publishing, Auburn, NH, 2011).

_____, 2011c, *From Asynchronous Logic to The Standard Model to Superflight to the Stars* (Blaha Research, Auburn, NH, 2011).

_____, 2012a, *From Asynchronous Logic to The Standard Model to Superflight to the Stars volume 2: Superluminal CP and CPT, U(4) Complex General Relativity and The Standard Model, Complex Vierbein General Relativity, Kinetic Theory, Thermodynamics* (Blaha Research, Auburn, NH, 2012).

_____, 2012b, *Standard Model Symmetries, And Four And Sixteen Dimension Complex Relativity; The Origin Of Higgs Mass Terms* (Blaha Reasearch, Auburn, NH, 2012).

_____, 2013a, *Multi-Stage Space Guns, Micro-Pulse Nuclear Rockets, and Faster-Than-Light Quark-Gluon Ion Drive Starships* (Blaha Research, Auburn, NH, 2013).

_____, 2013b, *The Bridge to Dark Matter; A New Sister Universe; Dark Energy; Inflatons; Quantum Big Bang; Superluminal Physics; An Extended Standard Model Based on Geometry* (Blaha Reasearch, Auburn, NH, 2013).

_____, 2014a, *Universes and Megaverses: From a New Standard Model to a Physical Megaverse; The Big Bang; Our Sister Universe's Wormhole; Origin of the Cosmological Constant, Spatial Asymmetry of the Universe, and its Web of Galaxies; A Baryonic Field between Universes and Particles; Megaverse Extended Wheeler-DeWitt Equation* (Blaha Reasearch, Auburn, NH, 2014).

_____, 2014b, *All the Megaverse! Starships Exploring the Endless Universes of the Cosmos Using the Baryonic Force* (Blaha Research, Auburn, NH, 2014).

_____, 2014c, *All the Megaverse! II Between Megaverse Universes: Quantum Entanglement Explained by the Megaverse Coherent Baryonic Radiation Devices – PHASERs Neutron Star Megaverse Slingshot Dynamics Spiritual and UFO Events, and the Megaverse Microscopic Entry into the Megaverse* (Blaha Research, Auburn, NH, 2014).

_____, 2015a, *PHYSICS IS LOGIC PAINTED ON THE VOID: Origin of Bare Masses and The Standard Model in Logic, U(4) Origin of the Generations, Normal and Dark Baryonic Forces, Dark Matter, Dark Energy, The Big Bang, Complex General Relativity, A Megaverse of Universe Particles* (Blaha Research, Auburn, NH, 2015).

_____, 2015b, *PHYSICS IS LOGIC Part II: The Theory of Everything, The Megaverse Theory of Everything, U(4)⊗U(4) Grand Unified Theory (GUT), Inertial Mass = Gravitational Mass, Unified Extended Standard Model and a New Complex General Relativity with Higgs Particles, Generation Group Higgs Particles* (Blaha Research, Auburn, NH, 2015).

_____, 2015c, *The Origin of Higgs ("God") Particles and the Higgs Mechanism: Physics is Logic III, Beyond Higgs – A Revamped Theory With a Local Arrow of Time, The Theory of Everything Enhanced, Why Inertial Frames are Special, Universes of the Mind* (Blaha Research, Auburn, NH, 2015).

_____, 2015d, *The Origin of the Eight Coupling Constants of The Theory of Everything: U(8) Grand Unified Theory of Everything (GUTE), S^8 Coupling Constant Symmetry, Space-Time Dependent Coupling Constants, Big Bang Vacuum Coupling Constants, Physics is Logic IV* (Blaha Research, Auburn, NH, 2015).

_____, 2016a, *New Types of Dark Matter, Big Bang Equipartition, and A New U(4) Symmetry in the Theory of Everything: Equipartition Principle for Fermions, Matter is 83.33% Dark, Penetrating the Veil of the Big Bang, Explicit QFT Quark Confinement and Charmonium, Physics is Logic V* (Blaha Research, Auburn, NH, 2016).

_____, 2016b, *The Periodic Table of the 192 Quarks and Leptons in The Theory of Everything: The U(4) Layer Group, Physics is Logic VI* (Blaha Research, Auburn, NH, 2016).

_____, 2016c, *New Boson Quantum Field Theory, Dark Matter Dynamics, Dark Matter Fermion Layer Mixing, Genesis of Higgs Particles, New Layer Higgs Masses, Higgs Coupling Constants, Non-Abelian Higgs Gauge Fields, Physics is Logic VII* (Blaha Research, Auburn, NH, 2016).

_____, 2016d, *Unification of the Strong Interactions and Gravitation: Quark Confinement Linked to Modified Short-Distance Gravity; Physics is Logic VIII* (Blaha Research, Auburn, NH, 2016).

_____, 2016e, *MoND: Unification of the Strong Interactions and Gravitation II, Quark Confinement Linked to Large-Scale Gravity, Physics is Logic IX* (Blaha Research, Auburn, NH, 2016).

_____, 2016f, *CQ Mechanics: A Unification of Quantum & Classical Mechanics, Quantum/Semi-Classical Entanglement, Quantum/Classical Path Integrals, Quantum/Classical Chaos* (Blaha Research, Auburn, NH, 2016).

_____, 2016g, *GEMS: Unified Gravity, ElectroMagnetic and Strong Interactions: Manifest Quark Confinement, A Solution for the Proton Spin Puzzle, Modified Gravity on the Galactic Scale* (Pingree Hill Publishing, Auburn, NH, 2016).

_____, 2016h, *Unification of the Seven Boson Interactions based on the Riemann-Christoffel Curvature Tensor* (Pingree Hill Publishing, Auburn, NH, 2016).

_____, 2017a, *Unification of the Eleven Boson Interactions based on 'Rotations of Interactions'* (Pingree Hill Publishing, Auburn, NH, 2017).

_____, 2017b, *The Origin of Fermions and Bosons, and Their Unification* (Pingree Hill Publishing, Auburn, NH, 2017).

_____, 2017c, *Megaverse: The Universe of Universes* (Pingree Hill Publishing, Auburn, NH, 2017).

_____, 2017d, *SuperSymmetry and the Unified SuperStandard Model* (Pingree Hill Publishing, Auburn, NH, 2017).

_____, 2017e, *From Qubits to the Unified SuperStandard Model with Embedded SuperStrings: A Derivation* (Pingree Hill Publishing, Auburn, NH, 2017).

_____, 2017f, *The Unified SuperStandard Model in Our Universe and the Megaverse: Quarks, ... ,* (Pingree Hill Publishing, Auburn, NH, 2017).

_____, 2018a, *The Unified SuperStandard Model and the Megaverse SECOND EDITION A Deeper Theory based on a New Particle Functional Space that Explicates Quantum Entanglement Spookiness (Volume 1)* (Pingree Hill Publishing, Auburn, NH, 2018).

_____, 2018b, *Cosmos Creation: The Unified SuperStandard Model, Volume 2, SECOND EDITION* (Pingree Hill Publishing, Auburn, NH, 2018).

_____, 2018c, *God Theory* (Pingree Hill Publishing, Auburn, NH, 2018).

_____, 2018d, *Immortal Eye: God Theory: Second Edition* (Pingree Hill Publishing, Auburn, NH, 2018).

_____, 2018e, *Unification of God Theory and Unified SuperStandard Model THIRD EDITION* (Pingree Hill Publishing, Auburn, NH, 2018).

_____, 2019a, *Calculation of: QED α = 1/137, and Other Coupling Constants of the Unified SuperStandard Theory* (Pingree Hill Publishing, Auburn, NH, 2019).

_____, 2019b, *Coupling Constants of the Unified SuperStandard Theory SECOND EDITION* (Pingree Hill Publishing, Auburn, NH, 2019).

_____, 2019c, *New Hybrid Quantum Big_Bang–Megaverse_Driven Universe with a Finite Big Bang and an Increasing Hubble Constant* (Pingree Hill Publishing, Auburn, NH, 2019).

_____, 2019d, *The Universe, The Electron and The Vacuum* (Pingree Hill Publishing, Auburn, NH, 2019).

_____, 2019e, *Quantum Big Bang – Quantum Vacuum Universes (Particles)* (Pingree Hill Publishing, Auburn, NH, 2019).

_____, 2019f, *The Exact QED Calculation of the Fine Structure Constant Implies ALL 4D Universes have the Same Physics/Life Prospects* (Pingree Hill Publishing, Auburn, NH, 2019).

_____, 2019g, *Unified SuperStandard Theory and the SuperUniverse Model: The Foundation of Science* (Pingree Hill Publishing, Auburn, NH, 2019).

_____, 2020a, *Quaternion Unified SuperStandard Theory (The QUeST) and Megaverse Octonion SuperStandard Theory (MOST)* (Pingree Hill Publishing, Auburn, NH, 2020).

_____, 2020b, *United Universes Quaternion Universe - Octonion Megaverse* (Pingree Hill Publishing, Auburn, NH, 2020).

_____, 2020c, *Unified SuperStandard Theories for Quaternion Universes & The Octonion Megaverse* (Pingree Hill Publishing, Auburn, NH, 2020).

_____, 2020d, *The Essence of Eternity: Quaternion & Octonion SuperStandard Theories* (Pingree Hill Publishing, Auburn, NH, 2020).

_____, 2020e, *The Essence of Eternity II* (Pingree Hill Publishing, Auburn, NH, 2020).

_____, 2020f, *A Very Conscious Universe* (Pingree Hill Publishing, Auburn, NH, 2020).

_____, 2020g, *Hypercomplex Universe* (Pingree Hill Publishing, Auburn, NH, 2020).

_____, 2020h, *Beneath the Quaternion Universe* (Pingree Hill Publishing, Auburn, NH, 2020).

_____, 2020i, *Why is the Universe Real? From Quaternion & Octonion to Real Coordinates* (Pingree Hill Publishing, Auburn, NH, 2020).

_____, 2020j, *The Origin of Universes: of Quaternion Unified SuperStandard Theory (QUeST); and of the Octonion Megaverse (UTMOST)* (Pingree Hill Publishing, Auburn, NH, 2020).

_____, 2020k, *The Seven Spaces of Creation: Octonion Cosmology* (Pingree Hill Publishing, Auburn, NH, 2020).

_____, 2020l, *From Octonion Cosmology to the Unified SuperStandard Theory of Particles* (Pingree Hill Publishing, Auburn, NH, 2020).

_____, 2021a, *Pioneering the Cosmos* (Pingree Hill Publishing, Auburn, NH, 2021).

_____, 2021b, *Pioneering the Cosmos II* (Pingree Hill Publishing, Auburn, NH, 2021).

_____, 2021c, *Beyond Octonion Cosmology* (Pingree Hill Publishing, Auburn, NH, 2021).

_____, 2021d, *Universes are Particles* (Pingree Hill Publishing, Auburn, NH, 2021).

_____, 2021e, *Octonion-like dna-based life, Universe expansion is decay, Emerging New Physics* (Pingree Hill Publishing, Auburn, NH, 2021).

_____, 2021f, *The Science of Creation New Quantum Field Theory of Spaces* (Pingree Hill Publishing, Auburn, NH, 2021).

_____, 2021g, *Quantum Space Theory With Application to Octonion Cosmology & Possibly To Fermionic Condensed Matter* (Pingree Hill Publishing, Auburn, NH, 2021).

68   **REFERENCES**

_____, 2021h, *21ˢᵗ Century Natural Philosophy of Octonion Cosmology , and Predestination, Fate, and Free Will* (Pingree Hill Publishing, Auburn, NH, 2021).

_____, 2021i, *Beyond Octonion Cosmology II : Origin of the Quantum; A New Generalized Field Theory (GiFT); A Proof of the Spectrum of Universes; Atoms in Higher Universes* (Pingree Hill Publishing, Auburn, NH, 2021).

_____, 2021j, *Integration of General Relativity and Quantum Theory: Octonion Cosmology, GiFT, Creation/Annihilation Spaces CASe, Reduction of Spaces to a Few Fermions and Symmetries in Fundamental Frames* (Pingree Hill Publishing, Auburn, NH, 2021).

_____, 2022, *New View: Hypercomplex Cosmology Based on the Unification of General Relativity and Quantum Theory* (Pingree Hill Publishing, Auburn, NH, 2022).

Eddington, A. S., 1952, *The Mathematical Theory of Relativity* (Cambridge University Press, Cambridge, U.K., 1952).

Fant, Karl M., 2005, *Logically Determined Design: Clockless System Design With NULL Convention Logic* (John Wiley and Sons, Hoboken, NJ, 2005).

Feinberg, G. and Shapiro, R., 1980, *Life Beyond Earth: The Intelligent Earthlings Guide to Life in the Universe* (William Morrow and Company, New York, 1980).

Gelfand, I. M., Fomin, S. V., Silverman, R. A. (tr), 2000, *Calculus of Variations* (Dover Publications, Mineola, NY, 2000).

Giaquinta, M., Modica, G., Souchek, J., 1998, *Cartesian Coordinates in the Calculus of Variations* Volumes I and II (Springer-Verlag, New York, 1998).

Giaquinta, M., Hildebrandt, S., 1996, *Calculus of Variations* Volumes I and II (Springer-Verlag, New York, 1996).

Gradshteyn, I. S. and Ryzhik, I. M., 1965, *Table of Integrals, Series, and Products* (Academic Press, New York, 1965).

Heitler, W., 1954, *The Quantum Theory of Radiation* (Claendon Press, Oxford, UK, 1954).

Huang, Kerson, 1992, *Quarks, Leptons & Gauge Fields 2ⁿᵈ Edition* (World Scientific Publishing Company, Singapore, 1992).

Jost, J., Li-Jost, X., 1998, *Calculus of Variations* (Cambridge University Press, New York, 1998).

Kaku, Michio, 1993, *Quantum Field Theory*, (Oxford University Press, New York, 1993).

Kirk, G. S. and Raven, J. E., 1962, *The Presocratic Philosophers* (Cambridge University Press, New York, 1962).

Landau, L. D. and Lifshitz, E. M., 1987, *Fluid Mechanics 2ⁿᵈ Edition*, (Pergamon Press, Elmsford, NY, 1987).

Misner, C. W., Thorne, K. S., and Wheeler, J. A., 1973, *Gravitation* (W. H. Freeman, New York, 1973).

Rescher, N., 1967, *The Philosophy of Leibniz* (Prentice-Hall, Englewood Cliffs, NJ, 1967).

Rieffel, Eleanor and Polak, Wolfgang, 2014, *Quantum Computing* (MIT Press, Cambridge, MA, 2014).

Riesz, Frigyes and Sz.-Nagy, Béla, 1990, *Functional Analysis* (Dover Publications, New York, 1990).

Sagan, H., 1993, *Introduction to the Calculus of Variations* (Dover Publications, Mineola, NY, 1993).

Sakurai, J. J., 1964, *Invariance Principles and Elementary Particles* (Princeton University Press, Princeton, NJ, 1964).

Weinberg, S., 1972, *Gravitation and Cosmology* (John Wiley and Sons, New York, 1972).

Weinberg, S., 1995, *The Quantum Theory of Fields Volume I* (Cambridge University Press, New York, 1995).

INDEX

About the Author

Stephen Blaha is a well-known Physicist and Man of Letters with interests in Science, Society and civilization, the Arts, and Technology. He had an Alfred P. Sloan Foundation scholarship in college. He received his Ph.D. in Physics from Rockefeller University. He has served on the faculties of several major universities. He was also a Member of the Technical Staff at Bell Laboratories, a manager at the Boston Globe Newspaper, a Director at Wang Laboratories, and President of Blaha Software Inc. and of Janus Associates Inc. (NH).

Among other achievements he was a co-discoverer of the "r potential" for heavy quark binding developing the first (and still the only demonstrable) non-Aeolian gauge theory with an "r" potential; first suggested the existence of topological structures in superfluid He-3; first proposed Yang-Mills theories would appear in condensed matter phenomena with non-scalar order parameters; first developed a grammar-based formalism for quantum computers and applied it to elementary particle theories; first developed a new form of quantum field theory without divergences (thus solving a major 60 year old problem that enabled a unified theory of the Standard Model and Quantum Gravity without divergences to be developed); first developed a formulation of complex General Relativity based on analytic continuation from real space-time; first developed a generalized non-homogeneous Robertson-Walker metric that enabled a quantum theory of the Big Bang to be developed without singularities at t = 0; first generalized Cauchy's theorem and Gauss' theorem to complex, curved multi-dimensional spaces; received Honorable Mention in the Gravity Research Foundation Essay Competition in 1978; first developed a physically acceptable theory of faster-than-light particles; first derived a composition of extremums method in the Calculus of Variations; first quantitatively suggested that inflationary periods in the history of the universe were not needed; first proved Gödel's Theorem implies Nature must be quantum; provided a new alternative to the Higgs Mechanism, and Higgs particles, to generate masses; first showed how to resolve logical paradoxes including Gödel's Undecidability Theorem by developing Operator Logic and Quantum Operator Logic; first developed a quantitative harmonic oscillator-like model of the life cycle, and interactions, of civilizations; first showed how equations describing superorganisms also apply to civilizations. A recent book shows his theory applies successfully to the past 14 years of history and to *new* archaeological data on Andean and Mayan civilizations as well as Early Anatolian and Egyptian civilizations.

He first developed an axiomatic derivation of the form of The Standard Model from geometry – space-time properties – The Unified SuperStandard Model. It unifies all the known forces of Nature. It also has a Dark Matter sector that includes a Dark ElectroWeak sector with Dark doublets and Dark gauge interactions. It uses quantum coordinates to remove infinities that crop up in most

interacting quantum field theories and additionally to remove the infinities that appear in the Big Bang and generate inflationary growth of the universe. It shows gravity has a MOND-like form without sacrificing Newton's Laws. It relates the interactions of the MOND-like sector of gravity with the r-potential of Quark Confinement. The axioms of the theory lead to the question of their origin. We suggest in the preceding edition of this book it can be attributed to an entity with God-like properties. We explore these properties in "God Theory" and show they predict that the Cosmos exists forever although individual universes (or incarnations of our universe) "come and go." Several other important results emerge from God Theory such a functionally triune God. The Unified SuperStandard Theory has many other important parts described in the Current Edition of *The Unified SuperStandard Theory* and expanded in subsequent volumes.

Blaha has had a major impact on a succession of elementary particle theories: his Ph.D. thesis (1970), and papers, showed that quantum field theory calculations to all orders in ladder approximations could not give scaling deep inelastic electron-nucleon scattering. He later showed the eigenvalue equation for the fine structure constant α in Johnson-Baker-Willey QED had a zero at $\alpha = 1$ not 1/137 by solving the Schwinger-Dyson equations to all orders in an approximation that agreed with exact results to 4th order in α thus ending interest in this theory. In 1979 at Prof. Ken Johnson's (MIT) suggestion he calculated the proton-neutron mass difference in the MIT bag model and found the result had the wrong sign reducing interest in the bag model. These results all appear in Physical Review papers. In the 2000's he repeatedly pointed out the shortcomings of SuperString theory and showed that The Standard Model's form could be derived from space-time geometry by an extension of Lorentz transformations to faster than light transformations. This deeper space-time basis greatly increases the possibility that it is part of THE fundamental theory. Recently, Blaha showed that the Weak interactions differed significantly from the Strong, electromagnetic and gravitation interactions in important respects while these interactions had similar features, and suggested that ElectroWeak theory, which is essentially a glued union of the Weak interactions and Electromagnetism, possibly modulo unknown Higgs particle features, be replaced by a unified theory of the other interactions combined with a stand-alone Weak interaction theory. Blaha also showed that, if Charmonium calculations are taken seriously, the Strong interaction coupling constant is only a factor of five larger than the electromagnetic coupling constant, and thus Strong interaction perturbation theory would make sense and yield physically meaningful results.

In graduate school (1965-71) he wrote substantial papers in elementary particles and group theory: The Inelastic E- P Structure Functions in a Gluon Model. Phys. Lett. B40:501-502,1972; Deep-Inelastic E-P Structure Functions In A Ladder Model With Spin 1/2 Nucleons, Phys.Rev. D3:510-523,1971; Continuum Contributions To The Pion Radius, Phys. Rev. 178:2167-2169,1969; Character Analysis of U(N) and SU(N), J. Math. Phys. 10, 2156 (1969); and The Calculation of the Irreducible Characters of the Symmetric Group in Terms of the

Compound Characters, (Published as Blaha's Lemma in D. E. Knuth's book: *The Art of Computer Programming Vols. 1 – 4*).

In the early 1980's Blaha was also a pioneer in the development of UNIX for financial, scientific and Internet applications: benchmarked UNIX versions showing that block size was critical for UNIX performance, developing financial modeling software, starting database benchmarking comparison studies, developing Internet-like UNIX networking (1982) and developing a hybrid shell programming technique (1982) that was a precursor to the PERL programming language. He was also the manager of the AT&T ten-year future products development database. His work helped lead to commercial UNIX on computers such as Sun Micros, IBM AIX minis, and Apple computers.

In the 1980's he pioneered the development of PC Desktop Publishing on laser printers and was nominated for three "Awards for Technical Excellence" in 1987 by PC Magazine for PC software products that he designed and developed.

Recently he has developed a theory of Megaverses – actual universes of which our universe is one – with quantum particle-like properties based on the Wheeler-DeWitt equation of Quantum Gravity. He has developed a theory of a baryonic force, which had been conjectured many years ago, and estimated the strength of the force based on discrepancies in measurements of the gravitational constant G. This force, operative in D-dimensional space, can be used to escape from our universe in "uniships" which are the equivalent of the faster-than-light starships proposed in the author's earlier books. Thus travel to other universes, as well as to other stars is possible.

Blaha also considered the complexified Wheeler-DeWitt equation and showed that its limitation to real-valued coordinates and metrics generated a Cosmological Constant in the Einstein equations.

Blaha has been developing hypercomplex cosmologies from January, 2020 to the present. He has published his work in a series of books.

The author has also recently written a series of books on the serious problems of the United States and their solution as well as a book on the decline of Mankind that will follow from current social and genetic trends in Mankind.

In the past twenty years Dr. Blaha has written over 80 books on a wide range of topics. Some recent major works are: *From Asynchronous Logic to The Standard Model to Superflight to the Stars*, *All the Universe!*, *SuperCivilizations: Civilizations as Superorganisms*, *America's Future: an Islamic Surge, ISIS, al Qaeda, World Epidemics, Ukraine, Russia-China Pact, US Leadership Crisis*, *The Rises and Falls of Man – Destiny – 3000 AD: New Support for a Superorganism MACRO-THEORY of CIVILIZATIONS From CURRENT WORLD TRENDS and NEW Peruvian, Pre-Mayan, Mayan, Anatolian, and Early Egyptian Data, with a Projection to 3000 AD*, and *Mankind in Decline: Genetic Disasters, Human-Animal Hybrids, Overpopulation, Pollution, Global Warming, Food and Water Shortages, Desertification, Poverty, Rising Violence, Genocide, Epidemics, Wars, Leadership Failure.*

He has taught approximately 4,000 students in undergraduate, graduate, and postgraduate corporate education courses primarily in major universities, and large companies and government agencies.

Recently he developed a quantum theory, The Unified SuperStandard Theory (UST), which describes elementary particles in detail without the difficulties of conventional quantum field theory. He found that the internal symmetries of this theory could be exactly derived from an octonion theory called QUeST. He further found that another octonion theory (UTMOST) describes the Megaverse. It can hold QUeST universes such as our own universe. It has an internal symmetry structure which is a superset of the QUeST internal symmetries.